An Illustrated Guide to

Animal Science
Terminology

WITH INTERACTIVE CD-ROM

Join us on the web at

agriculture.delmar.cengage.com

An Illustrated Guide to
Animal Science
Terminology

WITH INTERACTIVE CD-ROM

Colleen Brady

DELMAR
CENGAGE Learning™

Australia • Brazil • Japan • Korea • Mexico • Singapore • Spain • United Kingdom • United States

DELMAR
CENGAGE Learning

An Illustrated Guide to Animal Science Terminology with Interactive CD-ROM
Colleen Brady

Vice President, Career Education
 Strategic Business Unit: Dawn Gerrain

Director of Learning Solutions:
 John Fedor

Managing Editor: Robert L. Serenka, Jr.

Acquisitions Editor: David Rosenbaum

Product Manager: Christina Gifford

Editorial Assistant: Scott Royael

Director of Content & Media Production:
 Wendy A. Troeger

Production Manager: Mark Bernard

Content Project Manager: Jeffrey Varecka

Technology Project Manager:
 Sandy Charette

Director of Marketing: Wendy E. Mapstone

Marketing Manager: Gerard McAvey

Marketing Coordinator: Jonathan Sheehan

Art Director: David Arsenault

Cover Design: Studio Montage

For product information and technology assistance, contact us at
Cengage Learning Customer & Sales Support, 1-800-354-9706

For permission to use material from this text or product,
submit all requests online at **www.cengage.com/permissions**
Further permissions questions can be emailed to
permissionrequest@cengage.com

ISBN-13: 978-1-4180-1151-2

ISBN-10: 1-4180-1151-7

Delmar
Executive Woods
5 Maxwell Drive
Clifton Park, NY 12065
USA

Cengage Learning is a leading provider of customized learning solutions with office locations around the globe, including Singapore, the United Kingdom, Australia, Mexico, Brazil, and Japan. Locate your local office at **www.cengage.com/global**

Cengage Learning products are represented in Canada by Nelson Education, Ltd.

To learn more about Delmar, visit **www.cengage.com/delmar**

Purchase any of our products at your local bookstore or at our preferred online store **www.cengagebrain.com**

Notice to the Reader

Printed in the United States of America
2 3 4 5 6 7 15 14 13 12 11

Table of Contents

Preface

Animals have played a vital role in human existence since before recorded history. Over the millennia, a unique vocabulary has developed around the world of animal science. A comprehensive understanding of animal science terminology is vital for those wishing to seek a career in animal agriculture, and also important for the general public, as they are asked to make decisions and have opinions on issues in the field of animal agriculture and animal science. The purpose of this text is to provide a comprehensive list of animal science vocabulary, and to help the reader develop a basic understanding of the field of animal science.

This textbook consists of eighteen chapters that can be broken up into three distinct areas. The first part is an overview of common topics taught in introductory animal science courses. The second part focuses on species specific to production animal agriculture. The third part focuses on companion animal species.

In addition, the text includes a CD-ROM including all terminology with audio pronunciations.

HOW TO USE THIS TEXT

An Illustrated Guide to Animal Science Terminology helps you learn and retain terminology using a logical approach to word parts and associations. The keys to learning from this text include the following.

Terms

Terms appear in bold type, with the pronunciation and definition following.

Pronunciation System

The pronunciation system is an easy approach to learning the sounds of terms. This system is not laden with linguistic marks and variables so that the student does not get bogged down in understanding the key. Once students become familiar with the key it is easy for them to progress in speaking the language.

Pronunciation Key

Pronunciation guides for common words are omitted.

Any vowel that has a dash above it represents the long sound, as in

ā	hay
ē	we
ī	ice
ō	toe
ū	unicorn

Any vowel followed by an "h" represents the short sound, as in

ah	apple
eh	egg
ih	igloo
oh	pot
uh	cut

Unique letter combinations are as follows

oo	boot
ər	higher
oy	boy
aw	caught
ow	ouch

Other Pronunciation Guidelines

Word parts are represented in the text as prefixes, combining forms, and suffixes. The notion for a prefix is a word part followed by a hyphen. The notion for a combining form (word root and its vowel to ease pronunciation) is the root followed by a slash and its vowel, as in "nephr/o." The notion for a suffix is a hyphen followed by the word part. The terms prefix, combining form, and suffix will not appear in the definitions.

Learning Objectives

The beginning of each chapter lists learning objectives to tell students what is expected of them as they read the text and complete exercises.

Review Exercises

At the end of each chapter are exercises to help you interact with and review the chapter's information. The exercises include several formats: multiple choice, matching, case studies, word building, and diagram labeling. The answers to these exercises are found at the end of the book.

ABOUT THE AUTHOR

Colleen Brady grew up in rural Wisconsin, in a county with more Holstein dairy cattle than people. She quickly determined that a career in the field of animal sciences was the right choice for her. She began her college career at the University of Minnesota–Waseca, where she earned A.A.S. degrees in Animal Health Technology and Horse Management. After working in the Arabian horse industry for two years, she attended Michigan State University and earned B.S., M.S., and Ph.D. degrees in Animal Science. Dr. Brady is currently working at Purdue University, where she is an associate professor in the Department of Youth Development and Agricultural Education.

ACKNOWLEDGMENTS

The creation of a textbook is a task that no one completes alone. I would like to thank my colleagues in the College of Agriculture at Purdue, for their willingness to provide information that contributed to aspects of this text. Thanks also to my husband and friends, for their support and encouragement throughout the process. Finally, thanks to Chris Gifford, and the staff at Delmar for their support and understanding through the development of this text.

Special thanks to the following reviewers who provided valuable feedback during the development of this project:

Mary O'Horo Loomis, DVM
SUNY Canton, Canton, NY

Anthony Seykora, Ph.D.
University of Minnesota, St. Paul, MN

Rick Parker, Ph.D.
University of Idaho, Moscow, ID

Chapter 1
What Is Animal Science?

Chapter Objectives

▶ Define animal science

▶ Identify primary areas of animal science and what they address

▶ Describe the role of animal science in daily life

INTRODUCTION TO ANIMAL SCIENCE

Animals are a vital part of our culture and society, and have been for thousands of years. Animals are sources of food and fiber for homes and for the table, as well as integral parts of our households and recreational activities (see Table 1-1). In the United States, two out of three households have a pet in residence. Millions of people participate in exhibiting animals for pleasure and recreation (see Figures 1-1, 1-2, 1-3). A common misconception is that animal science is only for those interested in cattle, sheep, pigs, poultry, and horses. However, the field of animal science developed into the study of the biology and management of all domestic animals. In the last decades, animal scientists have become more and more concerned with issues relevant to not only animals that provide food and fiber, but also those that provide pleasure and companionship in our lives, such as dogs, cats, and companion birds.

The production of animals for human use, whether as food or as companions, is a multibillion-dollar annual industry in the United States, and is a vital part of our national economy. According to the United States Department of Agriculture (USDA), more than $70 billion of meat animals were produced in the

TABLE 1–1
Census of domestic animals in the United States

Birds (pet-type)[4]	10,100,000
Cattle and calves[1]	95,497,994
Cats[2]	77,600,000
Chickens (egg-laying)[1]	429,317,605
Chickens (meat-type)[1]	1,389,279,047
Dogs[2]	65,000,000
Ducks, Geese, Other[1]	Not available
Emus[1]	48,221
Horses[3]	9,200,000
Ostriches[1]	20,550
Pigs[1]	62,485,647
Turkeys[1]	93,028,191

1. 2002 National Agricultural Statistics Service survey.
2. American Pet Products Manufacturers Association 2003–2004 survey.
3. American Horse Council survey.
4. U.S. Pet Ownership and Demographics Sourcebook, 2001, American Veterinary Medical Association.

United States in 2005 (see Figure 1-4). The value of animals raised for meat has risen significantly over the last 40 years.

Animal science, and work conducted in animal science, affects everyone in the world. The care and production of animals became an important part of human societies worldwide when humans stopped the hunter-gatherer lifestyle, and adopted a more **agrarian**

1

FIGURE 1–1 Showing a Hereford cow (Courtesy USDA)

FIGURE 1–2 Child showing a Duroc hog (Courtesy of USDA)

FIGURE 1–3 Rounding up beef cattle on the range (Courtesy ARS)

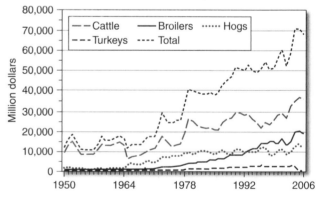

FIGURE 1–4 Total value of meat animals produced in the United States from 1948–2005 (Courtesy of USDA)

(ah-grahr-ē-ahn), or agriculture-based, village lifestyle. Subsequent to settling into villages, humans began the process of domesticating useful animals.

Domestic animals differ genetically from their wild ancestors, so the traits that are consistent with domestication are passed on to their offspring (see Table 1-2). These traits include changes in body size and shape, reduction of brain size, increased variety in color, increased deposit of fat and muscle, and retention of juvenile, or immature characteristics, which is known as **neoteny** (nē-oht-ah-nē). The process of domesticating animals is slow and occurs over many generations. The following are the stages of domestication:

1. Animals and humans began living in close proximity to one another in a symbiotic relationship. Animals and humans both benefited at this stage, but humans did not impose their will on the animals or deliberately provide food or shelter. The process of natural selection resulted in animals that were more comfortable in proximity to humans. These animals gained more benefit from the humans, and therefore, reproduced more successfully.

TABLE 1–2

The approximate years of domestication for common domestic animals in the United States

Species	Approximate year
Dog	14,000 B.C.
Sheep	8,000 B.C.
Goats/Pigs/Cattle	6,500 B.C.
Cats	6,000 B.C.
Llamas	4,000 B.C.
Donkeys/Horses	3,500 B.C.
Chickens	3,200 B.C.

FIGURE 1–5 A wolf in the wild. (Courtesy of USDA)

TABLE 1–3

U.S. per capita food consumption
Total meat: Boneless, trimmed (edible) weight, pounds per capita per year

Year	Red meat	Poultry	Fish and Shellfish	Total meat
2005	110	73.6	16.1	199.7
2004	112	72.7	16.5	201.2
2003	111.6	71.2	16.3	199.2
2002	114	70.7	15.6	200.4
2001	111.4	67.8	14.7	193.9
2000	113.7	67.9	15.2	196.8

Red meat commodities include, beef, veal, pork, lamb, and mutton. Poultry commodities include turkey and chicken, as well as the skin, neck, and giblets. Fish and shellfish include fresh and frozen, canned, and cured products. Game consumption for red meat, fishery products, and chicken for commercially prepared pet food are excluded. Numbers calculated from unrounded data.

Source: USDA/Economic Research Service. http://www.ers.usda.gov/Data/FoodConsumption/FoodAvailQueriable.aspx#midForm (last updated Dec. 21, 2005). U.S. Department of Commerce/National Marine Fisheries Service for fishery products.

2. Humans began confining animals and providing food and shelter, but not controlling breeding.
3. Humans began addressing selective breeding to emphasize desirable behavioral and physiological traits.
4. Humans continued selection and development of specific animal breeds that were genetically isolated from other animals in the species.

Different cultures domesticated different animals, depending on what was available and what their needs were. Several characteristics are common among animals that were domesticated. They generally were animals that lived comfortably in groups, and had some type of dominance hierarchy in place within their species group. They also tolerated humans, adapted to a variety of environments, and reproduced well in captivity.

It is important to note that taming is different than domestication, and is a behavioral change only, without the concurrent genetic changes that are present in domesticated animals. Animals that are tamed may allow more human interaction than nontamed animals, but the next generation will have the wild behavioral characteristics. Tamed animals also still have the full range of behavioral possibilities of their wild counterparts. This is important to remember when handling tamed animals, because they still have wild instincts and are different than the domestic species we work with regularly. Even the tamest of wolves is not a domestic dog (see Figure 1-5). It is also important to remember that offspring of domestic animals crossed with wild animals, such as wolf-dog hybrids, will display behaviors of both the wild and the domestic parent.

Humans found the effort of domestication worthwhile for many reasons. The first, and most obvious, was to increase access to a high-quality source of protein. Protein from a meat-based source is the most complete and bioavailable source of protein for people. Having domestic animals made that source of protein much more readily available and much more reliable than hunting. In 2004, Americans ate more than 200 pounds of meat per capita (see Table 1-3). In addition, humans benefited from the milk and egg production of the animals. Most domestic livestock were originally multipurpose animals. In modern animal science, we have developed breeds of animals that specialize in production of certain products that we use. Details on these breed differences are discussed in later chapters.

Domestic animals also served as guardians and in rodent control. As people began to grow plants and store crops from season to season, mice, rats, and other rodents were attracted to the ready supply of grain. Cats and dogs were especially useful in reducing these pest populations. One widely held theory about the beginning of domestication is actually built around this premise. Quite possibly, domestication of animals originated as dogs and cats were attracted to the readily available food supplies around human settlements. These food supplies include rodents that were attracted to the stored foodstuffs and refuse that included animal-based material from successful hunts.

Animals have also played a long and important role in transportation and as a source of power. Prior to the Industrial Revolution, animals were the primary source of power available to human societies. Horses, cattle, and buffalo have all served important roles as draft animals. Even breeds of dogs were developed to work as draft or pulling animals. In many countries, animals still play this important role. This is especially true in some Asian and African countries. Even in the United States, animals, especially horses, are used as a mode of transportation to access areas that cannot be reached with trucks or other vehicles.

The role of animals in our society has changed over the thousands of years of human-animal interaction, and continues to do so. In addition to serving roles as food sources, power sources, and for companionship, animals are playing an ever-increasing role in the health of humans (see Figure 1-6). Animals are very important in biomedical research, where they often serve as models for research that can improve the lives of people, and treat some of our most dreadful illnesses. Animals also serve an important role as service animals for many people with disabilities. These specially trained animals, usually dogs, provide support to help people with disabilities complete day-to-day tasks that many of us take for granted. Not only do these animals help with physical tasks, but they also facilitate social interactions between people. A growing area of research into the human-animal interaction addresses the role that animals can play in the development of young people. Animal interactions are believed to increase empathy in young people, as well as teach responsibility and many other important life skills.

As the roles of animals in our society change and evolve, the field of animal science has evolved to serve those changing needs. Animal science includes many specific areas of study. Nutrition, physiology, animal health, reproduction, and management are common

FIGURE 1–6 A person participating in a therapeutic riding program. (Courtesy Purdue University)

parts of most animal science programs and have been focus areas since the beginning of the study of animal science. Newer fields of study in animal science include ethology, biotechnology, and food safety.

COMMON TERMS IN ANIMAL SCIENCE

Animal agriculture   The segment of agriculture dealing directly with the production and use of animals. Animal agriculture generally refers to the production of those animals that are used for human consumption, whereas animal science refers to all domestic animals.

Animal health   The study of the overall condition and function of an animal, especially as related to disease. Animal health also includes the study of diseases that may impact people, whether through the food supply or through direct transmission of diseases from animals to people.

Animal rights   A philosophy that the rights of animals and humans are interchangeable. Supporters of animal rights do not believe animals should be used for research, animal agriculture, or as companions and for entertainment. (See animal welfare.)

Animal welfare   The philosophy that animals should be treated and managed with consideration for the animals' physical and psychological needs. Proponents of animal welfare focus on the humane use of animals by people, but do believe that using animals for research, food, and companionship is acceptable, as long as the animals are treated in a humane way.

Animal Welfare Act (AWA)   An act that regulates the care and treatment of animals. The AWA specifically addresses animal research facilities, animal dealers, animal exhibitors, and the transportation of animals. Components of the AWA also address animal fighting and pet theft.

Avian (ā-vē-ahn)   The species name for birds or bird-like animals.

Biotechnology (bī-ō-tehck-nohl-o-jē)   Any technique that uses living organisms to make or modify products, to improve plants or animals, or to develop microorganisms for specific purposes. Biotechnology is a growing field in animal science and agriculture. Biotechnology is used to increase reproductive efficiency of animals through techniques such as **cloning** (the production of an animal genetically identical to its parent). In 1996, the first livestock animal, a sheep, was cloned. Biotechnology is also used to assist animals in maximizing production. **Recombinant bovine somatotropin** (sō-mah-tō-trō-pihn), also known as rBST, has been used to supplement the naturally occurring **bovine somatotropin** (BST) in cattle to increase milk production. Some people resist the use of biotechnology in animals due to concerns about long-term affects. At the current time, the use of biotechnology is growing more rapidly in the plant sciences than in animal science.

Bovine (bō-vīn)   The species name for cattle.

Breed   A group of animals of the same species with a similar appearance and similar genetics that differentiate them from other animals of the same species.

Breed association   An organization that oversees the recording of registrations for a breed of animals. The breed association may also be involved in promoting the breed and hosting events, such as shows and sales, for the breed.

Breed character   Those physical characteristics (for example, size, color, coat pattern, coat type, etc.) that differentiate one breed from another.

Browse (browz)   Leaves and small twigs that animals consume. Goats and sheep are animals that commonly consume browse.

Canine (kā-nīn)   The species name for dogs.

Caprine (kahp-rīn)   The species name for goats.

Compost (kohm-pōst)   The material that results from decay of manure and other organic material. Compost can be used to fertilize fields and pastures, as well as gardens. Composted manure is available for residential use in many gardening stores.

Domestic animal (dō-mes-tihk)   An animal that has been bred over generations to benefit humans and to thrive in an environment of close human contact.

Draft animal   An animal used or developed for pulling carts, wagons, plows, or other items. Many developing countries still depend on draft animals as primary sources of labor. In the United States, draft animals are used by some religious groups, but are not a primary source of agricultural power.

Equine (ē-kwīn)   The species name for horses.

Ethology (ē-thohl-ō-jē)   The scientific study of the behavior of an animal, or group of animals, as related to the environment. In animal science, many ethology research programs focus on how the way we manage and care for our animals affects their behaviors and well-being. As people in animal production strive to meet consumers' demands to produce high-quality animal products, and to do so in a way that is not detrimental to the animal, the study of animal well-being is important.

Feline (fē-līn)   The species name for cats.

Feral (fehr-ehl)   An animal that was once domesticated and has returned to the wild state. The term *feral* is most often used to refer to horses and cats, but can apply to any animal. Feral animals are still genetically the same as their domestic counterparts. Mustangs of the western United States are feral horses.

Flock (flohck)   A group of birds or a group of sheep.

Flocking tendency   A desire in a group of sheep to gather together.

Food and Agricultural Organization of the United Nations (FAO)   The FAO is a part of the United Nations that promotes agricultural development around the world. The FAO also maintains statistics on agricultural production around the world.

Food safety   The study of conditions and practices that preserve the quality of food and minimize contamination and food-borne illness. The United States has the safest food supply in the world, and studies in food safety are designed to maintain and increase the level of safety. Food safety issues include bacterial contamination of food, especially during processing and handling, antibiotic residues in food products, and the role of genetic engineering in food products.

Generation (jehn-her-rā-shuhn)   A group of animals born around the same time, and that are approximately the same age.

Generation interval   The average time interval between the birth of parents and the birth of their offspring.

Genetics (jehn-eh-tihcks)   The science of how characteristics are passed on through generations.

Graze (grāz) To eat plant material that grows on the ground.

Grazing capacity The number of animals a particular piece of land can nutritionally support for a given time without damaging the land. Grazing capacity is influenced by the type of plant material present, the type of animals that are feeding on that plant material, and the typical environment in an area. The grazing capacity for a semiarid region is lower than the grazing capacity for a temperate region.

Grazing land The land that is used for grazing animals on a regular basis.

Grazing unit (1) The amount of pasture used by a mature cow, or its equivalent in another species, in a year; (2) A division of grazing land that is determined to assist in management of the property and livestock. A grazing unit can be of any size.

Grazing value The monetary value of the plant material on a grazing unit. This value is based on the palatability, nutritional value, quantity of forage, longevity of the plant material, and the area of distribution.

Gregariousness (greh-gahr-ē-ehs-nehs) The preference of some species of animals to remain in a group.

Heredity (heh-rehd-i-tē) The transfer of characteristics from parents to offspring.

Herd (herhd) A group of animals.

Herd book A record kept of the ancestry of an animal. Breed associations usually keep the official herd book for a breed.

Herdbound A reluctance or refusal to leave the herd.

Herding instinct In dogs, herding instinct is an instinctive desire to keep animals in groups, and to move those groups of animals together.

Mammal (mahm-al) Any of a group of warm-blooded animals that have fur and produce milk to feed their young.

Mammalian (mah-māyl-yuhn) Referring to mammals.

Management The direction or supervision of an animal science entity. Management could refer to a business or the direction of the care of animals.

Manure (mahn-oo-ər) The combination of fecal waste, urine, and bedding material produced by animals. Manure is commonly used as a fertilizer for fields and pastures. In some parts of the world, dried manure is burned as a source of fuel.

Neoteny (nē-oht-ah-nē) The retention of juvenile characteristics through adulthood. When compared to wild animals, adult domestic animals typically look more like the young in face and body shape. Adult domestic animals also retain more juvenile behaviors than adult wild animals.

Nutrition (noo-trish-uhn) The science of how a body uses food. The nutrients that animal bodies demand are the same nutrients that human bodies demand. There are many correlations between human nutrition and animal nutrition; in fact, animals are often used as models for studying issues such as diabetes and obesity relative to human nutrition. Nutritionists also explore how to use available foodstuffs to increase efficiency and well-being of animals. The largest economic segment of the companion animal industry is the pet food segment. As in human health and nutrition, animal nutrition increasingly focuses on how feeding and nutrition can be used to prevent disease and improve quality of life.

Organic agriculture (ōr-gahn-ihck) The raising of plants or animals for food with the minimal use of most conventional chemicals. The precise standards for labeling animal products as organic are available on the United States Department of Agriculture (USDA) Web site: www.USDA.gov. Organic agriculture is one of the fastest-growing segments of animal agriculture, as producers strive to meet the increasing demand for organic products.

Overgrazing A problem that occurs when too many animals are housed on too little land. The animals eat all of the plant material, and damage the root systems, resulting in severe damage to the grazing land. Overgrazed land can be restored, but significant resources are required to replant and fertilize the land.

Ovine (ō-vīn) The species name for sheep.

Pasture (pahs-chər) Land that is used for grazing animals.

Pasture rotation The process of rotating animals from one pasture to another through the growing season. Rotating pastures increases the efficiency of pasture use and decreases parasitic contamination of the pasture.

Physiology (fihz-ē-ohl-ō-jē) The science of the vital physical functions of living things. Physiologists study the whole animal, or study aspects of physiology at the cellular level. Physiology can be broken into several focused areas, including the following.

Exercise physiology A branch of physiology that primarily focuses on maximizing the athletic performance of dogs and horses. Physiologists also study the development of muscles, bones, tendons, and ligaments of animals, and seek ways to improve the quality of food products.

Growth physiology A field of physiology focused on maximizing growth and production of animals.

Lactation physiology Research conducted on the field of lactation and the mammary gland.

Reproductive physiology Research conducted on the reproductive processes of different animals, and the search for ways to improve reproductive efficiency and treat reproductive problems in animals.

Porcine (pōr'-sīn) The species name for pigs or hogs.

Range (rānj) (1) A term for land populated with native plants and grasses that is used for feeding animals; (2) The normal territory that a wild animal may inhabit.

Reproduction (rē-prō-duhck-shuhn) The science of how animals produce offspring. Research in reproduction has resulted in the use of the following technologies:

Artificial insemination The collection of semen from a male and the manual deposit of the semen in the reproductive tract of a female. The use of artificial insemination allows males to breed many more females than they could naturally. Artificial insemination also allows the shipment of semen, either cooled or frozen, to females around the world. Frozen semen keeps indefinitely, so semen can be preserved and used from males that have died. Artificial insemination also allows the breeding of males that may have an injury or other problem that makes it impossible for them to breed a female normally.

Embryo transfer A fertilized embryo taken from one female and implanted in another female to be carried through pregnancy. The use of embryo transfer technology allows genetically superior females to increase their contribution to the species. In some species, females can be **superovulated.** This can result in the release of numerous egg cells that can then be fertilized, with the embryos transferred to numerous other females.

Science The study of a subject in a planned and methodical way.

Species A group of living things with similar heritable characteristics.

Sustainable agriculture A practice of raising animals and crops that meets the current needs of society without reducing the potential for future production. Many of the concerns leading to the sustainable agriculture philosophy are based in concerns about the impact of high-intensity agriculture on the environment.

Trait A specific characteristic that can be passed through generations.

United States Department of Agriculture (USDA) The federal entity that regulates agriculture and the animal agriculture industry.

Xenotransplantation (zē-nō-trahnz-plahn-tā-shuhn) The transplantation of animal organs into humans.

CHAPTER SUMMARY

Animal science is a broad field that impacts the lives of people in a variety of ways. The production of food and fiber for our table and clothes, as well as a wide variety of other products (see Chapter 2) result from animal science. Although traditionally animal science has focused on production agriculture, in the last decade increasing interest and emphasis has been placed on the study of companion animals, such as dogs, cats, and birds. Animal science is also moving into fields such as ethology and biotechnology, and cooperative efforts between animal scientists and the medical profession are becoming more common as we seek ways to improve human and animal life.

STUDY QUESTIONS

1. Rank the species below in the order in which they were domesticated (1 is longest ago).

_____ Ovine   _____ Caprine   _____ Canine   _____ Feline

_____ Bovine   _____ Equine   _____ Porcine

2. Give one example of how biotechnology is currently applied that directly, or indirectly, affects animal agriculture.

3. List one aspect of animal science that has affected your life today.

4. Calculate the number of dogs and cats owned in your community.

   a) Calculate the number of households in your community by dividing the population of your community by the average number of people in a household (2.67 according to the 2000 U.S. Census).

   b) Multiply the estimation formula for the number of dogs (.58) by the total number of households.

   c) Multiply the estimation formula for the number of cats (.66) by the total number of households.

   Please note, these formulas are based on the national averages regarding number of households with pets, and are not designed to reflect the exact number of animals in your community. However, the calculation will provide a reasonable estimate of the number of dogs and cats in your community.

Below, match the animal science specialization with its description.

5. _____ Physiology

    a. The scientific study of animal behavior as it relates to its environment.

6. _____ Nutrition

    b. Any technique that uses living organisms to make or modify products, to improve plants or animals, or to develop microorganisms for specific purposes.

7. _____ Ethology

    c. The science of how food is used by the body.

8. _____ Biotechnology

    d. The direction or supervision of an animal science entity, which could be a business or the direction of the care of animals.

9. _____ Reproduction

    e. The science of the vital physical functions of living things.

10. _____ Management

    f. The science of how animals produce offspring.

Match the species name with the common name of these animals.

11. _____ Porcine

    a. Goat

12. _____ Canine

    b. Horse

13. _____ Feline

    c. Pig

14. _____ Bovine

    d. Sheep

15. _____ Equine

    e. Bird

16. _____ Ovine

    f. Dog

17. _____ Caprine

    g. Cat

18. _____ Avian

    h. Cow

Chapter 2
The Animal Agriculture Industry

Chapter Objectives

▶ Learn the diverse aspects of animal agriculture

▶ Learn the primary and secondary products of animal agriculture

▶ Learn the variety of careers in animal agriculture

SEGMENTS OF ANIMAL AGRICULTURE

Animal agriculture is often considered in terms of animal-raising farmers that live in rural area. However, the field of animal agriculture has a wide variety of types of operations and careers. People who raise animals are the foundations of the industry, but represent only a small percentage of the industry. In this chapter, we will define some of the diverse operations and opportunities in the animal agriculture industry.

Types of Animal Operations

Agri-tourism A segment of the animal agriculture industry that provides a recreational outlet for people. Agri-tourism operations range from dude ranches, where people can stay on a ranch, ride horses, and work with cattle, to petting farms and living history museums. At living history museums, people can immerse themselves in a historical time period and learn about life at that time.

Aquaculture A growing segment of the animal agriculture industry focuses on production of fish and

FIGURE 2–1 Fisherman with farm raised catfish (Courtesy of USDA)

other living water animals for consumption as food (see Figure 2-1). It is predicted that as more issues arise related to use of wild-caught fish in the diet, the field of aquaculture will continue to grow. In 2005, more than $1 billion of aquaculture products were sold in the United States, with the majority being food fish (see Figure 2-2). The aquaculture industry has the largest economic impact in Louisiana, Mississippi, Arkansas, and Alabama, but most states have some aquaculture industry (see Figure 2-3). As with many aspects of animal agriculture, the aquaculture industry has terms with unique definitions:

▶ **Hatchery** An operation that hatches fertilized eggs from breeding stock and then provides them to the nursery for raising.

9

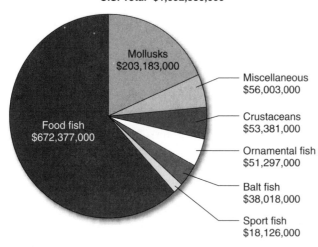

Value of Aquaculture Products Sold by Type: 2005
U.S. Total- $1,092,386,000

Mollusks
$203,183,000

Miscellaneous
$56,003,000

Crustaceans
$53,381,000

Food fish
$672,377,000

Ornamental fish
$51,297,000

Balt fish
$38,018,000

Sport fish
$18,126,000

FIGURE 2–2 Total value of all aquaculture products sold in the U.S. in 2005 (Courtesy of USDA)

▶ **Nursery** An operation that raises young fish to a size that is then moved into a finishing unit.
▶ **Pens** Net or cage enclosures that exist in large bodies of water, such as lakes, reservoirs, or the ocean.
▶ **Ponds** Solid enclosures created for raising animals.

Beef Cattle produced for their meat. The highest-quality cuts of meat in grocery stores and restaurants come from beef cattle. Many types of animal production operations are involved in beef cattle production, including the following:

▶ **Seedstock operation** An operation that produces cattle for breeding purposes. Most often, seedstock is purebred stock. Animals that are not of sufficient quality to be used for breeding animals are **culled,** or removed from the herd, and move into the market sector of the beef industry.

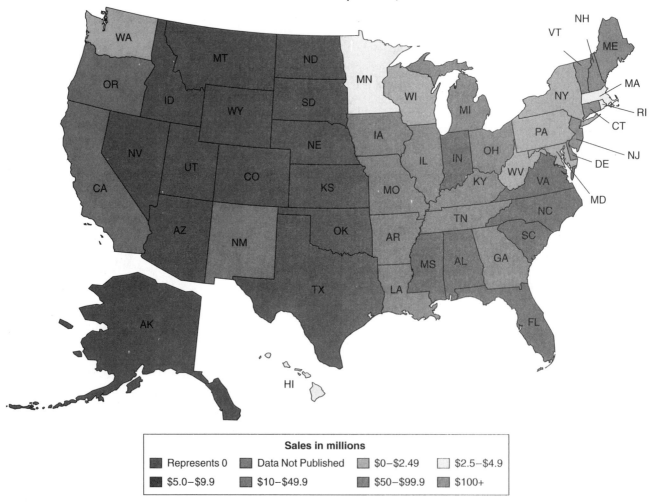

Aquaculture Sales: 2005
U.S. Total Sales - $1.09 Billion
Source: 2005 Census of Aquaculture, USDA-NASS

Sales in millions

| ■ Represents 0 | ■ Data Not Published | ■ $0–$2.49 | ☐ $2.5–$4.9 |
| ■ $5.0–$9.9 | ■ $10–$49.9 | ■ $50–$99.9 | ■ $100+ |

FIGURE 2–3 Value of aquaculture industry by state in 2005 (Courtesy of USDA)

FIGURE 2–4 Beef cow with nursing calf (Courtesy of Illinois at Urbana-Champaign)

▶ **Cow-calf operation** The primary focus for this operation is breeding cattle for market (see Figure 2-4). The majority of animals bred for market are crossbred animals. Calves will be raised until they are heavy enough to go to market. Calves that are underweight for the feedlot may be kept at the operation and raised until they are large enough,

or may be sold as **stocker calves,** which are calves that are purchased to be fed until they reach feedlot weight, and are then sold to a feedlot.

▶ **Feedlot operation** The primary focus of a feedlot is raising animals to market weight, or **finishing** the animals. Animals are between 650 and 800 pounds when they enter the feedlot (see Figure 2-5). There, they are raised to market weight, and then sent to market for processing. The animals are fed diets high in grain, resulting in the **grain-fed beef** seen in the grocery store.

▶ **Packing/processing plants** Facilities that slaughter animals when they reach their finished weight. Following slaughter, a United States Department of Agriculture (USDA) inspector examines the animals to ensure that they conform to all federal regulations (see Figure 2-6).

▶ **Vertical integration** This term describes a business with more than one type of operation, in which animals move from one operation to another with no change of ownership. For example, an owner of a cow-calf operation also owns a feedlot where the cattle go, or a packing/processing

FIGURE 2–5 Beef cattle in a feedlot

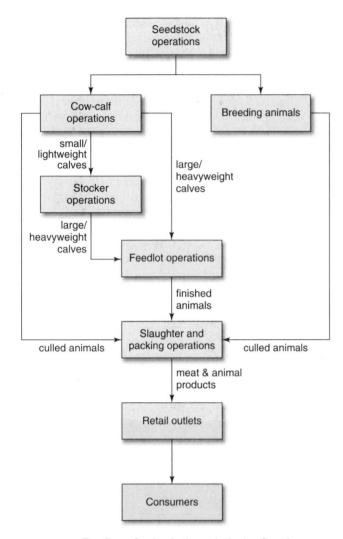

FIGURE 2–6 The flow of animals through the beef cattle industry (Source: Damron, 16-6)

FIGURE 2–7 Dog in a kennel

plant also owns a feedlot where the animals to be slaughtered came from.

Companion animals   Types of operations among companion animals vary little; therefore, they will be grouped here.

▶ **Boarding kennels** Operations that provide facilities for temporary housing of companion animals while their owners are away (see Figure 2-7). Boarding kennels may be associated with or operated independently of veterinary clinics. Some boarding kennels also offer "day care" services, where the animals are left at the kennel during the day while their owners are at work.

▶ **Breeders** People who raise companion animals for personal use or for sale. There are different types of breeders in the industry.

▶ **Hobbyist breeders** These breeders usually work with one or two breeds of animals. Hobbyist breeders are often very involved in exhibition of the offspring they produce in their breeding programs, and in selling their offspring to other hobbyist breeders. Many hobbyist breeders define some of their offspring as "pet quality," selling them to individuals for pets and requesting or requiring that they not be used for breeding. Most breeders of companion animals qualify as hobbyist breeders.

▶ **Puppy mills** This term is used to describe some breeding operations that produce large numbers of puppies under poor conditions. Puppy mills sell their puppies to brokers, who then sell them to retail sellers. The same principles can apply to breeders of other species who focus on producing maximum numbers of animals for resale, with minimal concern for the health and well-being of the animals, or for selection of quality animals as breeding stock.

▶ **Pet stores** Operations that purchase pets and pet-related products for resale.

Dairy   Cattle produced with a focus on milk production (see Figure 2-8). Dairy cows must have a calf every year to continue milk production. Some female calves are kept to become cows in the herd, and others are sold to different dairy farms or to feedlots, where they are raised to market weight and sent to market for processing.

▶ **Diversified farms** Traditional dairy farms that produce not only milk, but also may raise crops and other livestock. These dairy farms have fewer cattle (less than 200) than specialized farms where milk production is the only income generator.

FIGURE 2–10 Horse in a traditional stall at a boarding stable

FIGURE 2–8 Dairy farm (Courtesy ARS)

▶ **Drylot** A method of managing dairy cattle in which they live in large pens, and are provided all of their grains and forages. Drylots do not contain grass for grazing.

▶ **Free stalls** A housing arrangement for dairy cattle where stalls are provided, but the cattle may enter and exit the stalls at will (see Figure 2-9). Stall space is usually separated from feeding space.

FIGURE 2–9 Dairy cattle in a freestall housing system (Courtesy of ARS)

▶ **Specialized dairies** Dairy farms that concentrate on production of milk, and do not raise other livestock. Most specialized dairies purchase feed instead of raising the crops themselves.

Equine The equine industry uniquely straddles the traditional animal agriculture industries and the companion animal agriculture field. Horses have been an integral part of animal agriculture since they were first domesticated, and still serve important roles on some operations. The majority of horses in the United States are privately owned for recreational purposes, and a wide variety of horse operations provide services to these recreational owners.

▶ **Boarding operation** This service operation provides a facility where horses live and are cared for in exchange for a set fee (see Figure 2-10). The services and prices for boarding stables range tremendously. Some boarding stables have staff that provides riding instruction for people with horses at the stable.

▶ **Breeding operation** This operation can vary from one or two animals to hundreds of animals. The primary purpose of this type of operation is to produce offspring for sale. Most breeding operations produce purebred animals, but some operations produce animals for a specific purpose and are less concerned about the purity of lines. Breeding operations can be classified in two primary areas:

   ▸ **Mare station** A type of breeding operation where mares are bred and to deliver their foals. People who do not have the time or special skills needed to successfully breed their own mares use these operations.

FIGURE 2–11 Horse and rider jumping (Courtesy of Getty Images/PhotoDisc)

FIGURE 2–12 Chicks recently hatched at a hatchery (Courtesy ARS)

► **Stallion station** A type of breeding operation that stands several stallions to the public. These operations also frequently accept mares to be bred to any of their stallions, or mares to be bred to stallions from other places.

► **Private operation** Most horse operations are privately owned operations where horse owners provide care for horses they use for recreational purposes.

► **Training operation** This service operation provides racing, riding, or competition training for horses (see Figure 2-11). These trainers often show horses and provide riding lessons for their clients or the general public.

Poultry   The poultry industry has a diverse assortment of types of animal production facilities with different objectives. The poultry industry has a high degree of **vertical integration,** and most of the operations involved in production of poultry products are owned by one business entity.

► **Basic breeding** A facility housing the parent stock produces the fertilized eggs that are then incubated (see Figure 2-12).

► **Hatchery** Produces chicks or other baby birds. Hatcheries may produce birds for breeding, exhibition, egg production, or meat production (see Figure 2-13).

► **Grow-out farms** These facilities receive birds from hatcheries and either raise them to finished market weight in the case of meat-type birds, or raise them until they are ready to start laying eggs in the case of egg-type birds.

► **Processing plants** These plants process birds or eggs to prepare them for retail market. Refer to the products section of Chapter 11 for information on the wide range of poultry products available.

► **Value-added processing** Additional processing that increases the value of the product. For example, boneless, skinless chicken breasts are a value-added product of whole chickens.

Sheep and goats   The sheep and goat industries are relatively small, and serve primarily specialty markets. Although these industries are small in the United States, sheep and goats are major agricultural animals in other parts of the world, and were two of the first livestock species domesticated.

► **Dairies** Sheep and goats are raised around the world for milk production. The majority of sheep and goat milk produced in the United States is used for cheese production. These segments of the American animal agriculture are still very small (see Figure 2-14).

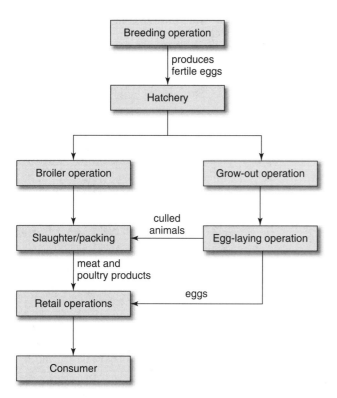

FIGURE 2–13 A schematic diagram of the vertical integration of the poultry industry. (Source: Damron, 3rd edition)

FIGURE 2–14 Milking a dairy goat (Courtesy Laurelwood Acres, CA)

▶ **Range production** Sheep and lambs are raised on pastures over large tracts of land. Lambs may go directly to **packing/processing plants,** or they may go to **feedlots.**

▶ **Secondary enterprise** Any enterprise where the income generated by the animals is not the primary source of income. Sheep and goats are often

raised as a secondary enterprise, or as part of a diversified farming plan. A secondary enterprise can also be an enterprise that provides income secondary to an off-farm enterprise, or job.

Swine The swine industry was traditionally located in the Midwest, where producers had easy access to the grains necessary for producing quality pork. However, over the last decades, the swine industry has grown significantly in the eastern part of the United States, and North Carolina is currently ranked second behind Iowa in total swine production (see Figure 2-15).

▶ **Seedstock operations** These operations produce purebred or crossbred offspring, either for use in breeding programs or for exhibition.

▶ **Farrow-to-finish** (fār-rō) An operation that breeds and raises pigs through market weight at the same location.

▶ **Feeder pig operation** An operation that raises pigs from weaning (separation from mother) until they are large enough to be finished for market.

▶ **Finishing operation** An operation similar to a feedlot in the beef industry. Finishing operations acquire feeder pigs, either by purchasing them or as part of a contract with another operation, and raise them to market weight.

▶ **Integrated production** A vertically integrated operation that runs operations at each level of production. An integrated production operation may or may not raise their own seedstock, but they will raise the pigs from birth to market.

Uses of Animals

The uses of animals is continually changing and adapting. Originally, their only role was to provide food; however, over thousands of years of animal-human interaction, animals have provided power, transportation, recreation, and companionship. Purposes of animal agriculture fit into several large categories as well. In this introductory section, we will discuss these categories in a general manner, with more detail found in each species chapter. In this section, we will consider contributions of companion animals that are not necessarily marketable.

By-products By-products are products that are left after the primary product has been removed. Leather from hides is a by-product, as is fat, which is used to make a variety of other products. Animal by-products are used to make common items, ranging from cosmetics to house insulation or human medicines. By-products can also be termed secondary products.

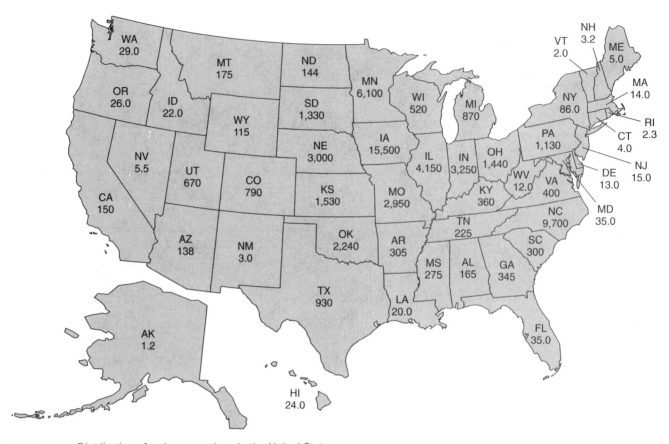

FIGURE 2–15 Distribution of swine operations in the United States

Fiber Animal hair or wool used to make cloth and other products.

Food The primary product from most of our traditional livestock species, hence many of them, especially cattle and swine, being known as "food animals." Food products take a variety of forms:

- ▶ **Meat** The tissue consisting of muscle and fat of animals that is processed for consumption.
- ▶ **Milk** The fluid produced by all mammals for nourishing their young. In the United States, cow's milk is the primary milk product, although sheep and goat's milk is used in the production of milk products such as cheese, butter, and yogurt. In other countries, mare's milk is also used for human consumption.
- ▶ **Edible by-products** Some items, such as certain organs like heart, kidney, and intestinal tissue, are edible and known as specialty foods. The by-products that are consumed are largely determined by culture and tradition.

Offspring Several species produce offspring as a product. For most companion animals, offspring is a potential income-generating product. In other species, offspring are produced either to enter the food-producing market, or as potential breeding stock. Dairy animals must produce offspring to produce milk. Many species have well-established exhibition traditions, and some animals are produced for the exhibition market (see Figure 2-16).

FIGURE 2–16 Youth showing lambs (Courtesy of USDA)

FIGURE 2–17 Youth showing a dog (Courtesy of USDA)

to reach youth at risk and help them to resolve their problems. Research in youth development is demonstrating that involvement with animals assists youth in developing lifelong skills.

CAREERS IN ANIMAL AGRICULTURE

With approximately 18 percent of the workforce in the United States employed in some aspect of animal agriculture, the broad field of animal agriculture has hundreds of careers that can be divided into the following five categories.

Farm production People working directly on the farm and with the animals. Farm production workers include farmers, managers, veterinarians, and so on.

Indirect employment Work in any variety of secondary industries that support animal agriculture, such as manufacturing of goods used in animal agriculture or produced by animal agriculture.

Processing and marketing The preparation and promotion of agricultural food and nonfood products for consumption.

Sales The sale of agricultural products to grocery stores, restaurants, and other retail markets, as well as wholesale markets. This also includes positions in the pharmaceutical industry providing products for farm production.

Suppliers People providing materials needed for farm production, such as machinery, chemicals, seed, and so on.

Recreation A primary outcome for companion animals and horses, recreation is a major economic generator across the country, and many animals are a part of these recreational activities. Recreation includes pleasure riding, as well as activities such as exhibition (see Figure 2-17), rodeos, racing, and so on.

Research The use of animals for research has contributed to tremendous progress in human health, and is partially credited for extending the human lifespan. In addition, animals are used for research to assist people in animal agriculture in more efficiently raising their animals and providing for better animal well-being in production animals. Through Institutional Animal Care and Use Committees, the federal government oversees animal research in the United States to ensure that animals are managed in compliance with the Animal Welfare Act.

Therapy A growing area of animal use is for therapy for people with physical, emotional, or cognitive disabilities. A number of programs are using animals

Specific Jobs in Animal Agriculture

Hundreds of types of jobs are available in animal agriculture, as well as opportunities for people to create specialty businesses that meet their needs. The following is a sampling of specific jobs in animal agriculture.

Agricultural economist A person who works within the agricultural industry to assess and impact the economic factors affecting agriculture.

Agriculture science and business (ASB) teacher A person who teaches courses in agricultural science and agricultural business to high school and/or junior high students. ASB teachers also frequently advise **Future Farmers of America (FFA) chapters,** an intracurricular youth organization designed to develop leadership skills and content knowledge in the field of agriculture.

Animal behaviorist A person specializing in studying and understanding animal behavior, and assisting in solving animal behavior problems.

Extension agent A person employed by a Cooperative Extension Service to provide an educational link between a land grant university and a community. Extension agents work with the 4-H youth program, which is the educational youth development arm of the USDA.

Farm broadcaster A person who reports farm-related news.

Farm manager A person who oversees the day-to-day running of an animal agriculture facility.

Field sales representative A person who works for a company, usually a feed company or a pharmaceutical company that markets products to wholesale or retail carriers.

Inspector Inspectors examine food products for safety and facilities to ensure that they are meeting federal and local standards for care and upkeep (see Figure 2-18). A variety of aspects of animal agriculture involve inspectors.

Laboratory technician A person working in a laboratory that has a relationship with the animal agriculture field. This could range from a person working with a veterinarian, to a researcher, to a person working in food science and safety.

Researcher A scientist who discovers new information in an area related to animal agriculture. Researchers can focus in any area that is related to animal agriculture.

FIGURE 2–18 Inspector at processing plant (Courtesy of USDA)

Veterinarian A veterinarian is a doctor who works with animals. The veterinary industry offers a variety of careers, ranging from veterinarians to kennel staff, receptionists, and office managers.

CHAPTER SUMMARY

Animal agriculture is a very diverse field, with a wide variety of types of operations and careers available. This chapter lists just some of the common production operations and career possibilities, but as science and technology continue to play an increasingly large role in animal agriculture, more opportunities will become available.

STUDY QUESTIONS

Match the aspect of the animal science industry with its definition.

1. _____ Aquaculture
2. _____ Boarding stable
3. _____ Stallion station
4. _____ Stocker
5. _____ Hatchery
6. _____ Vertical integration
7. _____ Feedlot
8. _____ Seedstock
9. _____ Puppy mill
10. _____ Kennel
11. _____ Farrow-to-finish
12. _____ Finishing operation

a. A place where beef cattle are fed to slaughter weight.
b. A place where chicks can be purchased.
c. Production of fish for human consumption.
d. A place where an owner can pay a fee to have a horse cared for.
e. A place where male horses are kept for breeding.
f. Calves that are postweaning, but too small for the feedlot.
g. A facility that houses dogs.
h. A swine facility that raises pigs from birth to market.
i. The ownership by one entity of all aspects of production.
j. A facility that feeds pigs to market weight.
k. Animals that are used for breeding.
l. An operation that raises large numbers of dogs in unacceptable conditions.

13. List the operations involved in beef cattle production, beginning with the birth of a calf and ending with the consumer purchase of a product.

14. What are the primary products of animal agriculture?

15. List ten careers in animal agriculture that are not listed in this chapter. Include the degree of education required or recommended for each position.

Chapter 3
Animal Genetics and Breeding

Chapter Objectives

► Learn how traits are inherited

► Learn the major criteria to consider in the selection of breeding stock

► Learn about heritable diseases

GENETICS

All animals pass on **traits** to their offspring through their **genes**. These genes contain the information necessary for the formation of a new animal. To make genetic progress, animals with the most desirable traits must be mated. **Selection** is the process of deciding which animals should be allowed to reproduce and pass their genes to the next generation. **Geneticists,** scientists who study genetics, can use mathematical calculations to predict the outcome of a mating, or a set of matings within a herd. Many genes influence some desirable characteristics, such as speed in a racehorse; fewer genes control other characteristics, such as animal height. The fewer the genes that control a characteristic, the easier it is to alter that characteristic through selective breeding. Use of technologies such as artificial insemination and embryo transfer allow animal scientists to increase the genetic contribution of individual males and females that have the most desirable genes. Commonly used terms related to genetics are:

Allele (uh-lēl)   A certain form of a specific gene that is located in a particular position on the chromosome. Alleles are labeled by use of one letter. If both alleles

on a chromosome are the same, they are indicated with the same letter (for example, BB or bb). If the alleles are different, they are labeled with upper- and lowercase letters (for example, Bb).

Aneuploid (ahn-ū-ployd)   An animal that has more or fewer chromosomes than an exact multiple of the monoploid number.

Animal breeding   The scientific application of principles of genetics to improve a species.

Artificial selection   The human selection of animals for reproducing based on the traits we consider most desirable. In artificial selection, humans, not the animals or other factors, control the opportunities for reproduction.

Autosomes (aw-tō-zōm)   All chromosomes other than sex chromosomes.

Breed   A group of animals that has been selected to have a specific set of characteristics that differentiates them from other members of the species. Breeds typically have visible qualities known as **breed characteristics** that are common among all members. Different breeds can be bred and result in viable offspring that are referred to as **crossbreds.**

Carrier   An animal that carries a gene but does not express the gene. This term is often used in relation to genetic diseases and defects.

Centromere (sehn-trō-mēr)   The part of a chromosome where the spindle fibers attach during division.

Chromosome (krō-mō-sōm)   A large molecule containing the genes of a living thing. Chromosomes are paired in most cells (see Table 3-1).

Codominance (kō-dohm-i-nahns)   When different alleles are located at one loci, but neither is dominant to the other. The phenotype reflects a combination of traits from both alleles, in a similar form to how they are seen in a homozygous state.

TABLE 3–1
The number of chromosomes in common animal species

Number of chromosomes by species	
Bovine	60
Canine	78
Caprine	60
Chicken	78
Equine	64
Feline	38
Ovine	54
Porcine	38

Crossbreeding   A form of outcrossing where animals of different breeds within a species are crossed. Crossbreeding maximizes **heterozygosity** in the offspring. This can be a benefit or a negative, depending on the goals of the breeder. For example, a dog that is the result of crossbreeding would be less likely to have a genetic disease associated with either parent; however, it would be less representative of the breed of either parent.

Deoxyribonucleic acid (DNA) (dē-ohck-sē-rī-bō-nū-klē-ihck ah-sihd)   The complex molecule that has all of the genetic information of a living thing. DNA is the "code" that determines all aspects of an organism's physical characteristics.

Dilution gene   A gene that affects the expression of another color gene. For example, in horses, the dilution of the chestnut coat color results in a palomino horse.

Diploid (dihp-loyd)   Cells containing two sets of chromosomes. Most cells are diploid cells.

Dominant (dohm-i-nahnt)   An allele that is expressed completely over the other allele in the pair, and completely masks the other allele. If an allele is dominant, it is labeled with an uppercase letter (for example, B).

Epistasis (eh-pihs-tah-sihs)   One gene or allele affecting the expression of another gene somewhere else on the chromosome.

Expression   The degree to which a particular gene, or set of genes, influences an animal. Some genes are **recessive,** and are not expressed at all. Some genes are **dominant,** and completely mask other genes.

Fancier (fahn-sē-ər)   A breeder who is interested in a specific breed of animal, and works toward the development and improvement of that breed. Often known by the species of the animal (for example, dog fancier, cat fancier, Angus fancier).

Foundation stock   An animal, or animals, identified as the genetic basis of the breed. The majority of animals in a breed will be descended from foundation stock.

Gamete (gahm-ēt)   Sex cells (sperm for males; ovum for females).

Gametogenesis (gahm-ē-tō-jehn-eh-sihs)   The process of forming gametes.

Gene (jēn)   A unit of a chromosome that determines the characteristics an animal may have. Genes affect not only the physical traits we see, but also the production of proteins that control all functions and activities in the body. In many cases, numerous genes work together to impact what we see as one trait. The more genes that control one trait, the more difficult to change that trait.

Gene frequency   How often a particular gene at a specific location occurs in a population. The population can be the population on a particular operation, or the population of a breed or species as a whole. Breeders that focus their matings on particular characteristics can increase the frequency of genes creating their desired result in their population of animals.

Gene pool   The full assortment of genetic options available for the breeder, relative to the genetic differences in animals, not the number of animals. In species that are endangered, the gene pool is quite small.

Gene splicing   Inserting new genetic material into a plasmid.

Gene transfer   Moving a gene from one animal to another.

Genetic drift   Changes in the genetic makeup of a population that are not the result of planned breeding, but are the result of random chance. For example, changes in the genes of a wild pig population would be due to genetic drift, since the breeding process is under no external control.

Genetic engineering   A broad term to describe a variety of methods of moving genes between animals.

Genetic trait summary   The ranking of beef sires based on the conformation of their female offspring.

Genome (jē-nōm)   All of the genetic information of a living thing. The genome provides a "map" for all the genes, but does not explain the role of each gene, or how the genes work together or affect each other. Mapping the genome is the first step in understanding the genetic makeup of a species.

TABLE 3–2
Color inheritance in horses

	B	b
B	BB	Bb
b	Bb	bb

Using a Punnett square, the percentage of offspring with a certain genotype can be determined from a mating. In the example above, each of the parents is phenotypically bay, with a genotype of Bb. (B is the bay gene, b is the chestnut gene.) Each parent contributes one color gene to the offspring. Genotypically, 25 percent of foals will be homozygous bay, 50 percent of the foals will be heterozygous bay, and 25 percent will be homozygous chestnut. Phenotypically, 75 percent of the offspring will be bay, and 25 percent will be chestnut.

Genotype (jē-nō-tīp)   The genes an animal possesses. These genes may or may not be expressed in their physical appearance (see Table 3-2).

Haploid (hahp-loyd)   Cells that contain a single set of chromosomes. Gametes (sex cells) are haploid cells. When gametes are joined, the resulting cell has two sets of chromosomes.

Heritability (hehr-ih-tah-bihl-ih-tē)   The amount of variation between individuals that is due to genetics.

Heritable trait (hehr-ih-tah-bihl)   A trait that can be passed from parent to offspring. Characteristics that are a result of the environment are not heritable, and cannot be passed onto the next generation.

Heterosis (heht-ehr-ō-sihs)   The increased health, growth, and performance that results from crossing animals with genotypes that are less similar than those found in the population (for example, by crossing animals of two different breeds). Also known as hybrid vigor.

Heterozygous (heht-ər-ō-zī-guhs)   When each allele, or gene, on the chromosome at a specific location is different. For example, an animal exhibiting Bb could pass either of two alleles/genes to its offspring.

Homologous chromosomes (hō-mohl-ō-guhs)   Chromosomes that are the same size and shape and affect the same traits.

Homozygous (hō-mō-zī-guhs)   When both alleles, or genes, on the chromosome at a specific location are the same. For example an animal that exhibits BB can pass only one allele/gene to its offspring.

Inbreeding   Mating closely related animals to concentrate a set of genes. The more closely related the animals, the more genes they will have in common. Inbreeding concentrates both desirable and undesirable traits.

Inbreeding depression   The decrease in health, growth, and performance that results from regular inbreeding. Inbreeding depression is a major concern in populations with a limited gene pool.

Incomplete dominance   A phenotype in which neither gene is completely dominant or completely recessive. The offspring will express a combination of the characteristics of each gene that is different than what either gene would express if it was homozygous.

Inheritance (in-hār-i-təns)   The passing of genes from one generation to the next.

Karyotype (kār-ē-ō-tīp)   A picture of all the chromosomes of an animal (see Figure 3-1). A karyotype is created by stopping mitosis, then staining the chromosomes with a dye. The different band colors in the photograph result because adenine and thymine absorb large amounts of the dye.

Linebreeding   A form of inbreeding in which animals with a shared ancestor are mated. That individual may be several generations removed, and may occur in the pedigree numerous times.

Locus (lō-kuhs)   The location of a gene on the chromosome. Plural form **loci** (lō-sī).

Meiosis (mī-ōh-sihs)   The process of cell division that results in four haploid cells, each containing one-half of the genetic material of the parent (see Figure 3-2). Meiosis is the cell division process by which gametes are formed.

Messenger ribonucleic acid (mRNA)   The template for the new protein. The mRNA communicates to the ribosome what amino acids are needed to build the specific protein.

Migration (mī-grā-shuhn)   The introduction of a new set of genes into a population. This can be achieved on a large scale or small scale by bringing a new male into the breeding program, or by a breed accepting crosses to another breed. For example, when the American Quarter Horse Association determined that Thoroughbreds could be bred into Quarter Horses, and produce registerable offspring, a new set of genes was introduced into the population.

Mitosis (mī-tō-sihs)   Normal cell division that results in a cell that is identical to the parent cell (see Figure 3-3). Mitosis is the cell division process by which somatic cells in the body divide for body growth or for repair of the body.

Modifier gene   A gene that has an effect on how another gene is expressed.

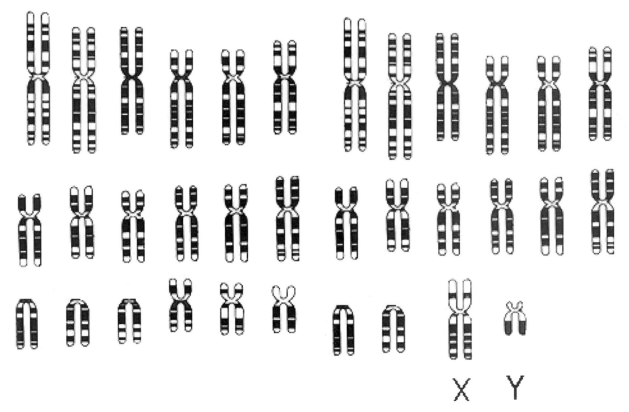

FIGURE 3–1 The karyotype of a horse

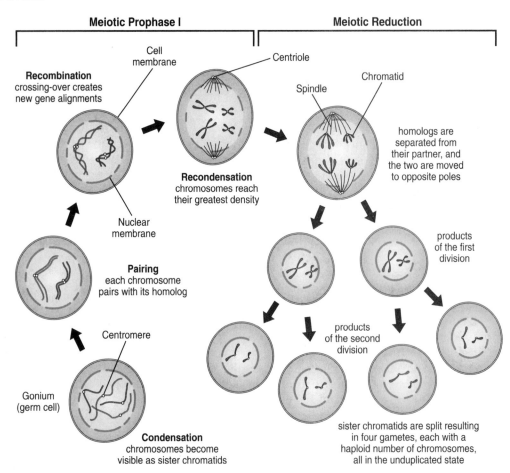

FIGURE 3–2 Cell division by meiosis, resulting in the formation of gametes.

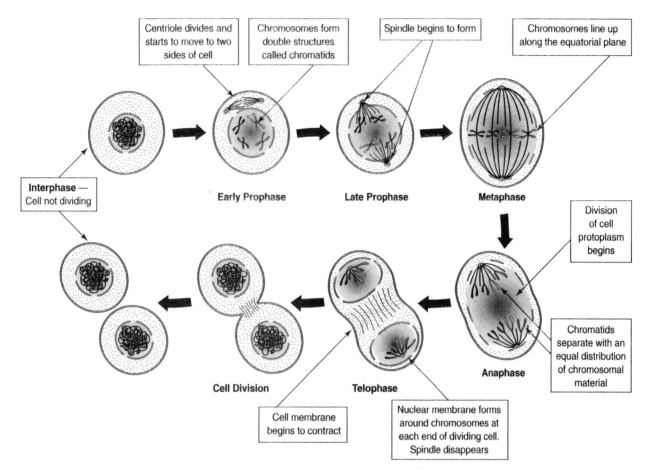

Early Prophase: Centriole divides and starts to move to two sides of cell; Chromosomes form double structures called chromatids

Late Prophase: Spindle begins to form

Metaphase: Chromosomes line up along the equatorial plane

Anaphase: Division of cell protoplasm begins; Chromatids separate with an equal distribution of chromosomal material

Telophase: Nuclear membrane forms around chromosomes at each end of dividing cell. Spindle disappears

Cell Division: Cell membrane begins to contract

Interphase — Cell not dividing

FIGURE 3–3 Cell division by mitosis, resulting in the formation of an identical cells. Mitosis occurs in somatic cells.

Multiple alleles The existence of more than two alleles for a given gene. Although only two alleles will exist in any given individual, there may be more than two alleles in the population of the species.

Mutagen (myoo-tah-jehn) Something that causes a change in the DNA that is permanent and transmissible to offspring. Mutagens can be chemical, physical, or radioactive in nature.

Mutation (myoo-tā-shuhn) A chemical change in the DNA of a gene, resulting in a new gene or allele. Mutations seldom occur in a population, but can result in rapid changes in the population when they do occur. Mutations can result in either positive or negative traits.

Natural selection Reproductive selection without human interference. Natural conditions determine which animals have the greatest opportunity to reproduce. The concept of "survival of the fittest" is based on natural selection.

Outcrossing Breeding of animals that are less related than the general population. Outcrossing is the opposite of inbreeding and increases the heterozygosity of the offspring.

Ovum (ō-vuhm) The female gamete, which carries the DNA from the female parent. Plural is **ova** (ō-vah).

Phenotype (fē-nō-tīp) The physical expression of genes in an animal. The phenotype is not always an accurate indicator of the genotype. The phenotype does not indicate which genes might be recessive, or which genes may be expressed in an incomplete way.

Plasmid (plahz-mihd) A portion of DNA in a cell that is not part of the chromosomal DNA, and that can reproduce itself.

Polyploid (pohl-ē-ployd) An animal that has more than two full sets of chromosomes.

Population genetics The study of frequencies of genes in a population, and how that frequency changes over time. Population genetics focuses on the group of animals, not on the traits of individual specimens.

Qualitative traits Traits that are categorized by their presence or absence (for example, coat color or presence of horns). These traits are usually controlled by a small number of genes.

Quantitative traits Traits that can be measured, such as production of milk or eggs, weight gain, racing speed, and so on. These traits are often controlled by many genes.

Recessive allele (rē-sehs-ihv) An allele in a pair that is not expressed in the presence of a dominant allele.

For a recessive allele to be expressed, both alleles in the pair must be recessive. Recessive alleles are labeled with a lowercase letter (for example, b).

Ribonucleic acid (RNA) (rī-bō-nū-klē-ihck ah-sihd) Molecules that are built as a complement to DNA; RNA carries the information to the cell to produce a specific protein.

Ribosome (rī-boh-zōhm) The part of the cell that manufactures proteins.

Sex chromosomes Chromosomes that determine the sex of the offspring. In mammals, females are XX and males are XY. Each parent passes on one sex chromosome to the offspring.

Sex-influenced traits Traits that are expressed differently in males and females with the same genotype.

Sex-limited traits Both males and females carry the genes for the traits, but the trait is only expressed in one sex. Milk production is an example: Males cannot express the trait; however, males from lines of females that are high producers are more likely to have female offspring that are high producers.

Sex-linked inheritance Genes that are located on parts of the sex chromosomes that are different. Therefore, the gene is only transferred to offspring that inherits the sex chromosome with that gene present.

Somatic cell (sō-maht-ihck) Any cell that is not a gamete.

Sperm A male gamete that carries the DNA of the male parent.

Strain A population within a breed of more closely related animals. When animals from different strains are crossed, the population is a **strain cross.**

Transcription The building of RNA to complement the DNA for a particular protein.

Transfer RNA (tRNA) The RNA molecules that carry the code that determines which amino acids are needed to build a protein, and recruit those proteins from the cytoplasm of the cell for the ribosomes to produce the required protein.

Transgenic (trahnz-gehn-ehk) Genetic material from another source has been introduced into the genetic material of a plant or animal. Transgenic mice with specific traits are often used for human medical research.

Translation The process of using the information from mRNA to assemble amino acids in the correct sequence for a particular protein.

Breeding/Mate Selection

Much thought should go into the decision to create offspring from two animals. The genetics of both animals must be considered to ensure a mating that accomplishes the intended purpose, whether for increasing production of meat or milk, matching a breed standard for a show dog or cat, or producing a fast racehorse. Most species have some diseases that are heritable within the species, or within certain lines in a species. Responsible breeders use caution to ensure that they are not perpetuating undesirable traits in the animals they breed. In some species, such as dogs and cats, overpopulation of animals is a significant issue. When people decide to breed dogs and cats, they should ensure that good homes are available for all of the offspring to ensure they are not contributing to the overpopulation problem. Many terms in animal science specifically relate to the mating and breeding of animals.

Breeding soundness examination A physical examination, usually by a veterinarian, to determine the ability of a male or female to reproduce. A breeding soundness exam of a male should include collection and evaluation of a semen sample. Breeding soundness exams of females should ensure that they are capable of conceiving and carrying a pregnancy to term. Breeding soundness examinations are especially important when purchasing new breeding stock.

Closed herd A herd of animals into which no new animals are introduced.

Congenital disease (cahn-jen-ə-təl) A disease or abnormality that has been present since birth, but is not heritable. A congenital disease can occur because of the environment during gestation or birth, and may or may not be due to a genetic change in the animal.

Dam The female parent of an animal.

Estimated Breeding Value (EBV) The value of an animal for breeding estimated by calculating performance of the individual and its relatives compared to the general population.

Expected Progeny Difference (EPD) A calculation that estimates how an individual's offspring will perform compared to other animals of the same generation.

Freemartin (frē-mar-tən) A female calf twin born with a male calf that has a high likelihood of infertility and underdeveloped reproductive organs.

Generation interval The average time between the birth of an animal and its production of offspring. Animals that reach puberty more quickly and have a shorter gestation time have a shorter generation interval than animals that mature more slowly and carry offspring longer.

Get Offspring of a male. Some shows have "Get of Sire" classes, where several offspring from one male are judged as a group.

Heritable disease A disease that is genetically inherited from one or more parent. An animal with a

```
                    ┌── Czort ──── Wielki Szlem
          ┌─ *El Paso ──┤              Forta
          │          └── Ellora ─── Witraz
Magic VF ─┤                          Elza
          │          ┌── *Bask ──── Witraz
          └─ Basks Maria┤             Balalajka
                      └── Judith-B ── Buszman
Czartonya D                           HMR Julie

          ┌── *El Paso ──┬── Czort ──── Wielki Szlem
          │              │              Forta
          │              └── Ellora ─── Witraz
          │                             Elza
Czartina ─┤
          │            ┌── Czortan ── Czort
          └── Czortina ─┤              Mortissa
                      └── *Gawra ─── Gerwazy
                                      Lafirynda
```

FIGURE 3–4 The pedigree of a horse with a high degree of inbreeding.

heritable disease has the genes to pass that disease on to its offspring, and use in breeding programs should be very limited. Heritable diseases exist in all species, and are inherited in a variety of ways. It is often desirable to test animals for heritable diseases prior to breeding.

Mate selection The process of identifying the most ideal animal to breed to a particular animal.

Pedigree (ped-ə-grē) The written record of an animal's ancestors. In a pedigree, the male ancestors are listed above the female ancestors for each generation (see Figure 3-4).

Produce The cumulative offspring of a female. Some shows have "Produce of Dam" classes, in which several offspring from one female are judged as a group.

Progeny (proh-jehn-ē) The cumulative offspring of a sire or dam.

Progeny testing The calculation of how the performance of offspring of a sire or dam compare to their contemporaries.

Sire (sīr) The male parent of an animal.

Sire summary Information provided by some breed associations, especially cattle, providing results from national evaluations of sires within a breed.

Animal Breeding

Many specific terms relate to the breeding of animals. Chapter 4 will address the physiology and anatomy of

the reproductive process. Chapter 6 will address behaviors associated with breeding and reproduction. This section focuses on terminology related to the reproductive management of animals. Reproduction is a vital aspect of animal science. Successful breeding is necessary to advance a species, and create a more desirable product, whether a meat animal or a companion animal that better serves the owner. The production of milk is not possible without the successful breeding of a female. Although we have made many changes in animals as we have domesticated them, the basics of the breeding process have remained the same. The gestation length for a domestic species is the same as for their analogous wild relatives. The success of breeding depends on having animals that are in good health and managers that are aware of the intricacies of the breeding process.

Terms for Male Animals

Barrow (bār-ō) A castrated male pig (see Figure 3-5).

Boar An intact male pig.

Buck An intact male goat or rabbit.

Bull An intact male bovine.

Capon (kā-pohn) A castrated male chicken.

Caponette (kā-pohn-eht) A male chicken that is sterilized with female hormones prior to sexual maturity.

Cockerel (kohck-ər-ahl) A young male chicken.

Colt A young male horse.

Dog An intact male canine.

Drake An intact male duck.

Gander (gahn-dehr) An intact male goose.

Gelding (gehl-dihng) A castrated male horse.

Gomer bull (gō-mər) A bull that is used to detect estrus in cattle. A gomer bull has been rendered infertile through vasectomization, sterilization, or surgical deviation of the penis to prevent breeding.

FIGURE 3–5 Youth showing a crossbred barrow at a hog show (Courtesy of USDA)

In the case of a castrated animal, the animal is treated with testosterone so it will exhibit sexual behavior.

Lapin (lah-pihn) A castrated male rabbit.

Ram An intact male sheep.

Rooster An intact male chicken.

Spotter bull A bull that has been vasectomized and is used to detect estrus in cows. Because he is vasectomized, not castrated, he still produces testosterone but cannot impregnate the cows.

Stallion (stahl-yuhn) An intact male horse.

Steer A castrated male bovine.

Tom An intact male cat or turkey.

Wether A castrated sheep or goat.

Additional Male-Related Terms

Cryptorchid (kriph-tōr-kihd) A mammal with one or both of the testicles remaining in the body cavity, and not descending into the scrotum. This condition may require surgical intervention. An animal with both testicles remaining in the body will be infertile, or subfertile, but may still produce male sex hormones and exhibit behaviors similar to those of an intact male.

Ejaculate (ē-jahck-yū-lāt) (verb) The release of semen from the reproductive tract of the male. Note the difference in the verb and noun forms of the same spelling.

Ejaculate (ē-jahck-yoo-lət) (noun) The amount of semen resulting from one ejaculation of the male. Note the difference in the verb and noun forms of the same spelling.

Erection The engorgement of the erectile tissue in the penis with blood, or through straightening of the **sigmoid flexure,** depending on the species.

Flehmen (fleh-mehn) The upward curling of the lip of a male of a species in response to scents detected from the female.

Semen (sē-mehn) The fluid males ejaculate that contains sperm cells and other fluids that nourish and protect the sperm cells, and create volume to facilitate the transport of sperm cells.

Sigmoid flexure (sihg-moyd flehck-shər) A bend in the penis that is extended in some species to achieve an erection

Terms for Female Animals and Offspring

Bitch A female dog.

Broodmare (brood-mār) A female horse whose primary purpose is producing foals.

Calf A young bovine of either gender.

Chick A young chicken of either gender; also used for most companion birds.

FIGURE 3–6 A beef cow and her calf (Courtesy of American Polled Hereford Association, MO)

Cow A female bovine (see Figure 3-6).

Doe A female goat or rabbit.

Duckling A young duck of either gender.

Ewe (ū) A female sheep.

Filly (fihl-ē) A young female horse.

Foal (fōl) A young horse of either gender.

Gilt A young female pig.

Goose A female goose.

Gosling A young goose of either gender.

Heifer (hehf-ər) A young female bovine.

Hen A female chicken, duck, or turkey.

Kid A young goat of either gender.

Kitten A young cat of any gender.

Lamb (lahm) A young sheep of either gender.

Mare A mature female horse.

Piglet A young pig of either gender.

Pullet (puhl-leht) A young female chicken.

Puppy A young dog of either gender.

Queen A female cat.

Sow A mature female pig or guinea pig.

Additional Female-Related Terms

Anestrus (ahn-ehs-truhs) An extended period of sexual inactivity in animals that are seasonally polyestrous or monestrous.

Corpus luteum (kōr-puhs loo-tē-uhm) The tissue formed after ovulation that eventually will produce progesterone, the hormone that is dominant when the female is pregnant or in diestrus. Corpus luteum is usually abbreviated as CL.

Estrous cycle (ehs-truhs) The pattern of hormonal and behavioral characteristics that mark the reproductive status of females (see Figure 3-7). Each stage of the estrous cycles has unique characteristics. The length of each stage of the cycle and the frequency of cycles vary among species. The following listing is in the order in which the stages of the cycle occur.

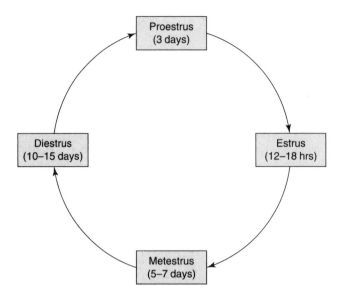

FIGURE 3–7 Estrous cycle of a cow

▶ **Proestrus** (prō-ehs-truhs) A period when follicle-stimulating hormone (FSH) is being released to stimulate follicular growth. The female is not sexually receptive.

▶ **Estrus** (ehs-truhs) The period of sexual receptivity. The primary hormone during this period is estrogen. Note that **estrus** is spelled differently when it refers to one stage of the **estrous** cycle.

▶ **Metestrus** (meht-ehs-truhs) A period following estrus that is characterized by the development of the corpus luteum and production of progesterone.

▶ **Diestrus** (dī-ehs-truhs) The period during which the animal's body determines pregnancy. During this time, progesterone is the primary hormone. If not pregnant, the animal continues to the proestrus period and back through the complete cycle. If pregnant, the animal continues through gestation.

▶ **Interestrous** (ihn-tər-ehs-truhs) A period in cats that occurs if ovulation is not induced via copulation.

▶ **Monestrous** (mohn-ehs-truhs) One estrous cycle each year, followed by a period of anestrus.

▶ **Polyestrous** (pohl-ē-ehs-truhs) Each estrous cycle is immediately followed by another estrous cycle.

▶ **Seasonally polyestrous** Each estrous cycle is followed immediately by another estrous cycle during part of the year. Then the female goes into anestrus for a period of time. Seasonally polyestrous animals can either be long-day breeders, in which the lengthening days stimulate a return to reproductive activity, or short-day breeders, in which shortening days stimulate the return to reproductive activity.

Gestation (jehs-tā-shuhn) In mammals, the period of time that the offspring develops in the uterus.

Gravid (grahv-ihd) Pregnant.

Incubation (ihn-kyoo-bā-shuhn) In avians, the length of time fertilized eggs need to be kept in a warm, humid environment to allow development of the young.

Ovulation (oh-vū-lā-shuhn) The release of the oocyte (or yolk in the case of birds) from the follicle. In most species, ovulation is spontaneous and occurs regardless of mating; however, some species (rabbits, cats, llamas) are **induced ovulators,** and must have the stimulation of the penis to induce the production of the hormone necessary for ovulation.

▶ **Ovulation fossa** (fah-sah) In horses, ovulation occurs through the center of the ovary. All other species ovulate on the edge of the ovary. The depression that is left when the mare ovulates is the ovulation fossa.

Ovulatory follicle (oh-vū-lah-tōr-ē) A follicle that has matured and is prepared to ovulate an ovum.

Parturition (pahr-tyoo-rihsh-uhn) The delivery of the offspring and the associated tissues and fluids. Also known as labor.

Primigravida (prih-mih-grahv-ih-dah) First pregnancy.

Primiparous (prih-mih-pāhr-uhs) A female that is pregnant for the first time, or has given birth to its first offspring.

Other Breeding-Related Terms

Artificial insemination (AI) AI is the insemination of collected semen into a female by a technician, instead of a male of the species (see Figure 3-8). This is a valuable reproductive management tool. AI allows for the insemination of numerous females with one ejaculate, and allows for preservation and transportation of semen. Precise techniques for collection and insemination vary with species.

Artificial vagina An instrument that imitates the vagina of a female and is used to collect semen from males for analysis or artificial insemination.

Atresia (ə-trē-szhə) The degeneration of follicles that do not mature to the ovulatory stage. Most follicles go through atresia.

Capacitation (kah-pahs-ih-tā-shuhn) A metabolic change in the sperm cell that gives it the ability to enter and fertilize an egg.

Castration (kahs-trā-shuhn) The removal of the testicles (see Figure 3-9). Castration results in sterility and removes the organs responsible for producing

FIGURE 3–8 Artificial insemination of a mare (Courtesy James Strawser, The University of Georgia)

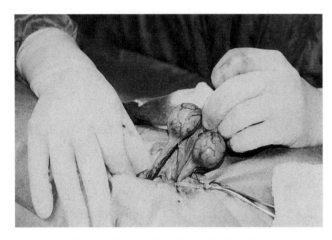

FIGURE 3–9 Surgical castration of a male dog. (Courtesy Lodi Veterinary Hospital, S.C.)

sex hormones in males. This removal is sometimes also called **neuter** (nū-tər), especially in companion animals.

Colostrum (kō-lah-struhm)   The first milk produced for the offspring. Colostrum is rich in antibodies and nutrients and is vital for the young to thrive.

Conception (kohn-sehp-shuhn)   The fertilization of the ovum by the sperm resulting in a zygote (zī-gōt).

Conception rate   The percentage of animals that get pregnant the first time they are bred in the breeding season. Usually used in relation to cattle.

Dystocia (dihs-tōs-ē-ah)   A difficult birth.

Electro ejaculation (ē-lehck-trō-ē-jahck-yoo-lā-shuhn)   Electrical stimulation of nerves in the male reproductive tract to cause ejaculation. This is a useful tool for collecting semen for analysis or for artificial insemination.

Embryo (ehm-brē-ō)   A developing zygote.

Fertilization (fər-tihl-ih-zā-shuhn)   The joining of the sperm and the ova.

Fetotomy (fē-toh-tō-mē)   The surgical cutting of the fetus to remove it from the uterus. This is done in the case when the fetus has died, and the female cannot deliver it normally.

Fetus (fē-tuhs)   Used to describe an unborn mammal, usually toward the end of gestation. Embryo is used as the descriptor earlier in gestation.

Gamete (gam-ēt)   A mature reproductive cell of either a male or a female.

Gonad (gō-nahd)   A general term for sex organs. Can refer to either male or female.

Implantation (ihm-plahn-tā-shuhn)   The attachment of a fertilized egg to the wall of the uterus. Implantation occurs prior to development of the placenta.

Mastitis (mahs-tī-this)   Inflammation of the mammary gland.

Morphology (mōr-fah-lō-jē)   The form and shape of the sperm cells. Sperm cells that are misshapen or malformed (see Figure 3-10) are less likely to be successful in fertilizing an egg.

Motility   The ability of a sperm cell to move progressively in a line. When semen is evaluated for quality, the **morphology, motility,** and **concentration** (number of sperm cells in a sample) are the primary criteria evaluated.

Oogenesis (ō-ō-jehn-eh-sihs)   The creation of egg cells.

Oophorectomy (ō-ōhf-ō-rehck-tō-mē)   The surgical removal of one or both ovaries. Only the ovaries are removed, leaving the rest of the reproductive tract.

Ovariohysterectomy (ō-vahr-ē-ō-hihs-tər-ehck-tō-mē)   The surgical removal of the ovaries, oviducts, and uterus in females. This procedure results in sterility, and the cessation of reproductive hormone production. Also known as **spay** (spā), especially in companion animals.

Ovigonium (ō-vih-gō-nē-uhm)   The germ cell from which the oocyte is produced.

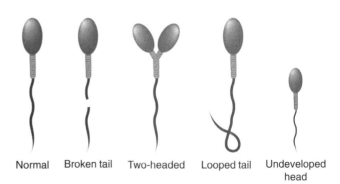

Normal   Broken tail   Two-headed   Looped tail   Undeveloped head

FIGURE 3–10 Morphology of sperm cells

Puberty (pū-behr-tē) The process through which an immature animal reaches sexual maturity. Age at puberty varies with species, and even in breeds within species. Environment and nutrition have a significant affect on the onset of puberty.

Retained placenta Occurs when the placenta, or parts of the placenta, remain attached to the uterus after parturition. The placenta may cause a significant uterine infection, and result in infertility if the condition is not treated promptly.

Spermatagonium (spər-mah-tō-gōn-ē-uhm) A primary germ cell in the testis of the male that produces the sperm cell.

Spermatogenesis (spər-mah-tō-jenn-eh-sihs) The creation of new sperm.

Urethra (yoo-rē-thrah) The tubular structure that carries urine out of the body. In males, the urethra also carries sperm from the testicles, through the penis.

Vasectomy (vah-sehk-tō-mē) The surgical removal of the vas deferens, which is the structure that carries sperm from the testis to the urethra. This results in sterility of the male because the sperm cannot travel from the testes; however, the male still produces sexual hormones and still exhibits male sexual behavior.

Zygote (zī-gōt) A fertilized egg.

CHAPTER SUMMARY

Breeding and genetics create the foundation for progress in animal species. Careful selection and mating of animals results in the creation of cattle and pigs that produce leaner meat, horses that run faster and jump higher, and dogs that more distinctively represent their breed. To make breeding decisions that move a breeding program in the desired direction, it is important to understand how traits are inherited, and how we can affect what traits are passed on.

Reproduction is a vital part of any animal agriculture effort. Without effective and efficient reproduction, we have no production of meat or milk. In companion animals, control of reproduction is a priority as we fight an ongoing battle with animal overpopulation and the propagation of genetic diseases in animals. A basic understanding of how characteristics are inherited, and the process involved in reproduction is vital for anyone involved in the animal industry.

STUDY QUESTIONS

1. Complete the following table with the correct information:

Species	Male Castrate	Male Intact
Bovine		
Canine		
Equine		
Chicken		

Match the stage of the estrous cycle with its description.

2. _____ Estrus

3. _____ Proestrus

4. _____ Metestrus

5. _____ Diestrus

6. _____ Anestrus

a. A period when no activity is occurring.

b. The period between estrus periods.

c. The time when FSH is stimulating follicular growth.

d. The period when the female is receptive to the male.

e. The time immediately after the cessation of receptive behavior.

7. Which of the following refers to the change in genes due to natural selection?

a. Population genetics

b. Genetic drift

c. Genetic waves

d. Gene frequency

8. Which describes the visual appearance of the animal?
   a. Phenotype
   b. Genotype
   c. Heterozygous
   d. Homozygous

9. Which breeding strategy results in animals that are the most similar genetically?
   a. crossbreeding
   b. natural selection
   c. inbreeding
   d. outcrossing

10. A _____ gene affects the expression of a color. This gene changes a _____ horse to a palomino.

11. A _____ allele is present in the animal, but is not expressed.

12. A male animal with one testicle that does not descend into the scrotum is a/an _____.

13. The surgical procedure to remove all reproductive organs of a female is a/an _____.

14. A horse breeder bred a bay stallion and a bay mare, and they had a chestnut offspring. In horses, the chestnut color is recessive. If bay is represented by B, and chestnut is represented by b, what is the genotype of the foal? What is the genotype of each parent? What percentage of the offspring of this mating would be bay? What percentage would be chestnut?

15. Animal species have a wide range of heritable diseases. Select the species of your choice, and conduct an Internet search to determine what heritable diseases are of concern in that species. Select one disease, research how it is inherited, and make a recommendation on how to control the disease in the population.

Chapter 4
Animal Anatomy

Chapter Objectives

► Identify the points of external anatomy of animal species

► Identify points of skeletal anatomy of animal species

► Understand types of connective tissue and how they differ

► Understand types of muscle and how they differ

► Know the parts and functions of major body systems

An understanding of external and internal anatomy is important in many fields of animal science. It is vital to understand where various cuts of meat originate, why animals have the characteristics they have, and how and why we manage animals in certain ways. Animal scientists in the field of physiology study various aspects of anatomy, depending on their specialty areas. **Exercise physiologists** may study skeletal, muscular, and respiratory anatomy, as they seek solutions to problems in athletic animals. **Muscle physiologists** study the skeleton and muscles as they try to solve muscular problems, as well as increase the amount and quality of muscle in meat animals.

SKELETAL ANATOMY

The skeleton is composed of various types of **connective tissue** that provide the structure and form of the body and bind the parts of the body together. When evaluating animals for different purposes, thorough knowledge of the skeleton is important. The alignment of bones, especially in the legs, is vital to the functionality and longevity of the animal. Skeletal structure is inherited, and structural defects can be passed to future generations. Therefore, animals that are more skeletally correct will remain in the milking or breeding herd for a longer period of time.

The **axial** (ăcks-ē-ahl) skeleton refers to the central skeleton that includes the skull, the vertebral column, and the ribs; the **appendicular** (ăhp-ən-dĭk-yoo-lər) skeleton refers to the limbs and appendages of the body. Differences in skeletal anatomy of avian species and mammals are relatively slight; therefore, differences between avian species and other mammals will be indicated in the definition of the skeletal part. The following are terms that relate to both the axial and appendicular skeletons:

Bone   Bone is the hardest connective tissue, and provides structure and form to the body. In birds, the bones are lighter weight and hollow, which allows some birds to fly.

Cartilage (cahr-tuh-ludge)   This connective tissue is more flexible than bone, and can be found either on parts of the bone or independent of bone. Ears are examples of cartilage that are independent of bone. The following are the different types of cartilage:

Articular (ahr-tik-yoo-lər)   Covers the surface of the bone in some types of joints.

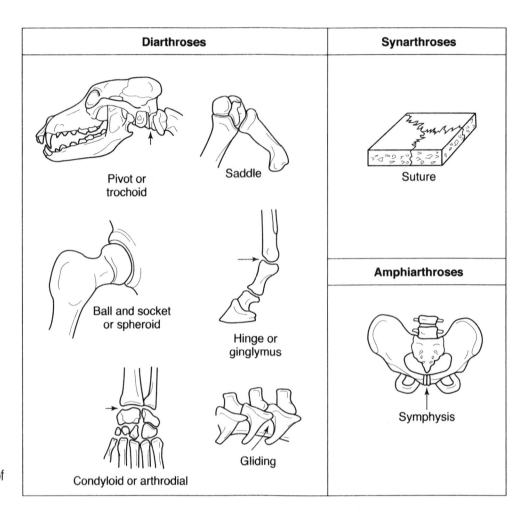

Diarthroses		Synarthroses

Pivot or trochoid

Saddle

Suture

Ball and socket or spheroid

Hinge or ginglymus

Amphiarthroses

Condyloid or arthrodial

Gliding

Symphysis

FIGURE 4–1 Different types of joints

Meniscus (meh-nis-kuhs)   Cartilage located between bone surfaces to cushion the surfaces (for example, stifle joint).

Joint   Where bones come together. Bodies have a variety of types of joints (see Figure 4-1). The following are the different types of joints:

Arthroidal (ahrth-roy-duhl)   A joint in which small oval-shaped bones fit into a depression in a large bone (for example, carpal joint).

Ball and socket   Highly movable joints in which the end of one bone is shaped into a ball that fits into a depression (socket) of the other bone in the joint (for example, hip joint).

Hinge   Joints that allow movement in only one direction, such as the stifle joint.

Suture (soo-chər)   An immovable joint where two bones join (for example, bones of the skull).

Symphysis (sihm-fih-sihs)   A type of joint where the two bones meet and work as a single bone, are joined by cartilage, and have limited movement (for example, **pelvis**).

Trochoid   Joints that operate with a pulley-type function, such as the atlas/axial joint.

Ligament (lihg-ah-mehnt)   Connective tissue that attaches bone to bone.

Tendon (tehn-dohn)   Connective tissue that attaches muscle to bone.

Axial Skeleton

The **axial** skeleton of avian species and mammals is very similar, so differences in the axial skeleton will be noted with each particular bone. The following are the components of the axial skeleton from the front of the body to the rear:

Skull (skuhl)   The skull is the bony structure that provides shape to the head. The following are the bones and regions that make up the skull:

Cranium (krā-nē-uhm)   The portion of the skull that surrounds and protects the brain.

Nasal (nā-sahl)   The bone that creates the bridge of the nose.

Nasal septum (sehp-tuhm)   The cartilage that divides the nasal cavity into two structures.

Sinuses (sīn-uhs-ehz)   Spaces in the skull that are filled with air or fluid.

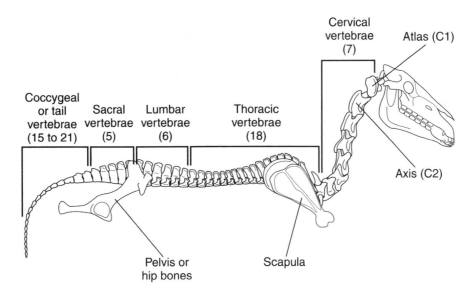

FIGURE 4–2 The appendicular skeleton of the horse

Vertebral column (vər-tē-brahl)   The column of bone also known as the backbone or the spine (see Figure 4-2). The vertebral column is made up of many individual **vertebra** (vər-tə-brah). The plural form of vertebra is **vertebrae** (vər-tə-brā). Vertebrae are grouped-based on their location in the vertebral column, and are numbered from the head to the tail. The following list of vertebrae is in order from the skull to the end of the spine:

Cervical (sihr-vih-kahl)   The vertebrae that constitute the neck. All mammals have seven cervical vertebrae, regardless of the length of their necks. Cervical vertebrae are numbered C1 through C7, with C1 being nearest the skull.

C1 is also called the **atlas,** and C2 is also called the **axis.**

Thoracic (thōr-ahs-ihck)   The vertebrae next to the cervical. Thoracic vertebrae form the external part of the **back** (or **chine**), and are attached to the **ribs.** You can locate the end of the thoracic vertebrae by locating the last rib.

Lumbar (luhm-bahr)   The vertebrae following the thoracic. Lumbar vertebrae do not have ribs attached, and are supported by muscle only. The lumbar region is also termed the **loin.**

Sacral (sā-krahl)   The vertebrae following the lumbar. These vertebrae make up the **croup** or **rump** (the external term varies with species).

TABLE 4–1
The number of vertebrae in common domestic species

Species	Cervical	Thoracic	Lumbar	Sacral	Coccygeal
Avian	14	7	Fused-14		6
Bovine	7	13	6	5	18–20
Canine/Feline	7	13	7	3	6–23
Caprine	7	13	6–7	4	16–18
Equine	7	18	6*	5	15–21
Ovine	7	13	6–7	4	16–18
Porcine	7	14–15	6–7	4	20–23

*Some Arabians have five lumbar vertebrae.

Coccygeal (kohck-sih-jē-ahl) Also known as **caudal** (kaw-duhl) vertebrae, the coccygeal vertebrae are the last in the vertebral column, and make up the **tail.**

Ribs The paired bones that attach on one side to the thoracic vertebrae, and on the other side to the **sternum** (stər-nuhm). The sternum is also known as the **breastbone** (or **keel** in avian species). **Barrel** is a common term used to describe the rib and body area.

Appendicular Skeleton

The **appendicular** skeleton includes the limbs of the body. In **quadrupeds** (kwa-droo-pĕdz)—four-legged animals—the limbs are commonly divided into the forequarter, which includes all parts of the shoulder and front leg, and the hindquarter (hīnd-kwar-tər), which includes all parts of the hip and hind leg (see Figure 4-3 and Figure 4-4).

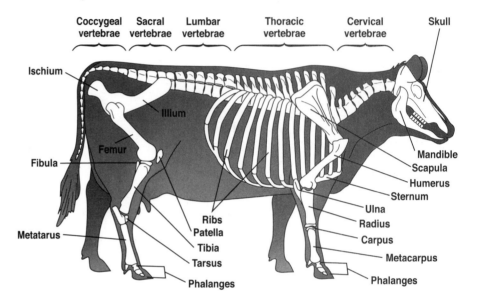

FIGURE 4–3 The skeleton of the cow

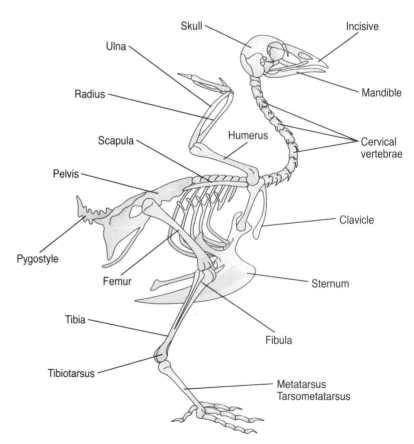

FIGURE 4–4 The skeleton of the chicken

Skeleton of the Forequarter

The forequarter carries the majority of the weight of quadrupeds. As an animal moves, a greater percentage of its weight is carried by the forequarter. When galloping, there is a point in the stride where the entire weight of the animal is on one foreleg. The following are the parts of the forequarter skeleton, from the top of the leg to the ground:

Scapula (skahp-yoo-lah)   Also the shoulder blade or **shoulder.** The scapula meets with the thoracic vertebrae to form the **withers.** To locate the scapula, use the point of the shoulder, where the scapula and humerus join, and the middle of the withers as landmarks.

Humerus (hū-mər-uhs)   The large bone in the forequarter that goes from the **point of the shoulder** to the **elbow.** The **arm** is the external equivalent.

Ulna and radius   The ulna (uhl-nah) and radius (rā-dē-uhs) are a pair of bones that form the largest long bone in the front leg. The top of the ulna creates the elbow. In some species, the ulna and radius are separate bones; in other species, the bones are fused. The external equivalent of the ulna and radius is the **forearm.**

Carpus (kahr-puhs)   Also known as the carpal joint, this is a multi-bone joint in the front leg of an animal, located below the ulna and radius. It is commonly known as the **knee** in large animals. The human equivalent is the wrist.

Metacarpus (meht-a-kahr-puhs)   The bones in the front leg below the carpus bones. Several metacarpal bones are in each leg, and they are numbered from the inside of the leg (**medial;** mē-dē-əhl) to the outside (**lateral**). In large animals, the third metacarpal bone is commonly referred to as the **cannon** bone.

Phalanges (fā-lahn-jēz)   The bones in the foot. Most species have multiple phalanges. Phalanges are numbered from closest to the body (**proximal**) to farthest from the body (**distal**), and are often coded with a P and the number. In large animals, P1 is the **long pastern,** P2 is the **short pastern,** and P3 is the **coffin bone** (see Figure 4-5). Phalanges are present in the forequarter and the hindquarter. Singular: **phalanx.**

Skeleton of the Hindquarter

The hindquarter begins at the pelvis and continues to the ground. Where the forequarter holds most of the weight, the hindquarter serves as the engine for moving the body. The hindquarter is more heavily muscled than the forequarter. In meat-producing animals, more desirable cuts of meat come from the hindquarter of

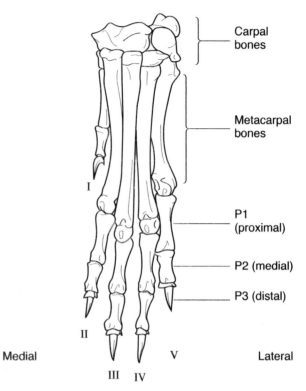

FIGURE 4–5 Numbering of phalanges and carpal bones. Phalanges are numbered proximally to distally. Carpal and tarsal bones are numbered medially to laterally.

the animal. In athletic animals such as horses and dogs, the skeleton of the hindquarter is of great importance in generating speed and impulsion in movement. The following are the parts of the hindquarter skeleton, from the top of the animal to the ground:

Pelvis (pehl-vihs)   The pelvis (hip) is made of three pairs of bones that act as one (see Figure 4-6). During parturition, the joints holding the bones of the pelvis relax, allowing expansion of the pelvis and facilitating the delivery of the offspring. One of the external signs of impending parturition is this relaxation of the pelvis. The following are the bones that make up the pelvis:

Acetabulum (ahs-eh-tahb-yoo-luhm)   The large round socket created where the three bones of the pelvis meet. The acetabulum is the "socket" of the ball and socket joint that creates the point of the hip.

Ilium (ihl-ē-uhm)   The largest of the bones of the pelvis, the ilium is shaped like a blade. In ruminant animals, the external part that is the ilium is known as the **hook.**

Ischium (ish-ē-uhm)   The ischium is smaller than the ilium, and forms the back of the pelvis, when viewed from the side. In ruminant animals, the external part of the anatomy that is the ischium is known as the **pins.**

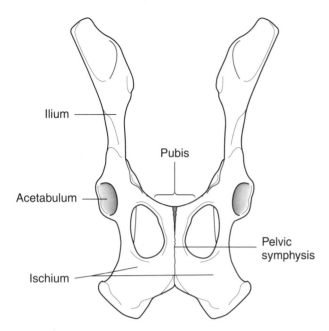

FIGURE 4–6 Parts of the pelvis

Pubis (pew-bihs) The pubis is the smallest bone of the pelvis, and is located on the bottom of the pelvis. The pubic bones are fused in a joint known as the **pubic symphysis** (pehw-bihck sihm-fih-sihs).

Femur (fē-muhr) The large bone in the hind leg. The top of the femur creates a **ball and socket joint,** the hip; the bottom of the femur creates a **hinge joint,** the **stifle** (stī-fuhl).

Patella (pah-tehl-ah) This is the equivalent to the kneecap in humans, and is a large sesamoid bone that is located between the femur and the tibia in the hindquarter. The term **knee** is used in quadrupeds to refer to the carpal joint in the fore-quarter, and is not used to refer to the patellar joint in the hindquarter. This joint is referred to as the **stifle.**

Tibia (tihb-ē-ah) and **fibula** (fihb-yoo-lah) Paired bones in the hind leg between the stifle and the **hock.** The tibia is large and carries the majority of the weight. The fibula is a slender bone that does minimal weight-bearing. In some species the fibula is fused to the tibia, and in others does not completely extend the length of the tibia.

Tarsus (tahr-suhs) The multi-bone joint in the hind leg below the tibia and fibula. The tarsus is also known as the hock in large animals. The tarsus is equivalent to the ankle in humans.

Metatarsus (meht-a-tahr-suhs) The bones in the hind leg below the tarsus. Just as with the metacarpus, there are several metatarsal bones, and they are numbered from the inside of the leg (medial) to the outside (lateral).

Sesamoid (sehs-ah-moyd) Small bones in both the front and hind legs. They are in the joint capsules, or embedded in tendons, in different locations. The patella is an example of a sesamoid bone, as is the **navicular** bone, which is in the hoof of the horse at P3.

External Anatomy

In addition to knowing the skeletal parts of the anatomy, it is important to know the common terms used to refer to the external parts of the animal that are associated with the skeletal areas. Terms for the same skeletal part often vary in different species. Knowing what terms are most appropriate for the specific species that is being discussed is also important. The following lists categorize those terms that are similar within species, as well as those that are different between species.

Avian Species

The basic external parts are the same for all avian species (see Figure 4-7). The following is a list of those parts:

Beak (bēk) The hard structure that surrounds the mouth of avian species. In some companion bird species, the beak must be kept trimmed to allow the bird to eat normally. Some poultry producers trim the beaks of (**debeak**) chickens to prevent the chickens from injuring each other through pecking.

Comb (kōm) The fleshy growth on the top of the head of a chicken. Males and females have combs, although the male comb is usually more prominent. Refer to Chapter 11 for details on different varieties of combs.

Feathers The outer covering of members of the avian species. Feathers are made of protein, and protect the body from the elements, much as hair does for mammals. The following are the many varieties of feathers:

▶ **Contour** The feathers covering the body of the bird.

▶ **Coverts** (cō-vərts) The feathers on the wings that start at the top of the wing and cover the bases of the primary and secondary feathers.

▶ **Down** Small soft feathers close to the body, or present on young. Down is one of the by-products of the poultry industry, and is used in pillows, comforters, and other warm clothing.

▶ **Hackles** Long, slender feathers on the neck of some birds.

▶ **Primary** The large feathers on the ends of the wings.

▶ **Secondary** The large, stiff feathers on the portion of the wing near a bird's body.

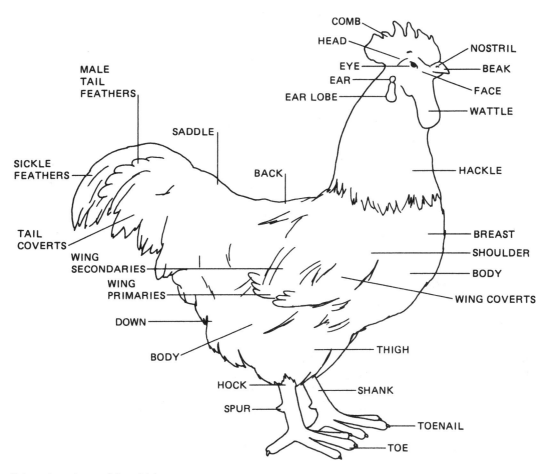

FIGURE 4–7 External anatomy of the chicken

Saddle The caudal portion of the topline of a bird.

Shank The leg of the chicken between the hock and the foot.

Spurs A sharp, bony projection from the shank of some types of poultry. The bird can use the spur as part of a defense, and can cause injury to other birds or people.

Wattle (what-tuhl) Fleshy tissue located near the chin in some poultry species that dangles down below the head.

Mammals

Most anatomy is the same in all four-legged animals (see Figure 4-8). Differences between species will be described in the figures. Many of the external parts are based on the location of external landmarks, which allow estimation of where the bones lie. Judging and evaluating many livestock species is a popular activity for youth and adults. A thorough understanding of the skeletal and external anatomy is vital for accurate evaluation of an animal, and clear communication of the results of that evaluation. The following are the external parts of mammals:

Cheek (chēk) The muscular region on the side of the face below the eyes and leading forward toward the lips.

Chest (chehst) A region in the front of the animal below the neck and between the points of both shoulders.

Coffin (kawf-ihn) **joint** The joint between P3, the coffin bone, and P2, the short pastern bone in hooved animals.

Coronet (kor-oh-net) Also known as the **coronary** (kor-ō-nār-ē) **band**, the coronet is the area where the skin of the leg meets the hoof (see Figure 4-9). Injuries to the coronary band can result in deformations of the hoof.

Crest (krehst) The top of the neck.

Croup (krewp) The region across the top of the hindquarter from the hips to the tail. Known as the **rump** in some species.

Dewclaw (doo-claw) A horny accessory digit on the back of the fetlock of many ruminant animals (see Figure 4-10) and pigs. In dogs and cats, dewclaws are the evolutionary remnants of toes. Some breeders choose to remove dewclaws at birth because they may become injured or damaged, especially in hunting dogs that work in fields and woods.

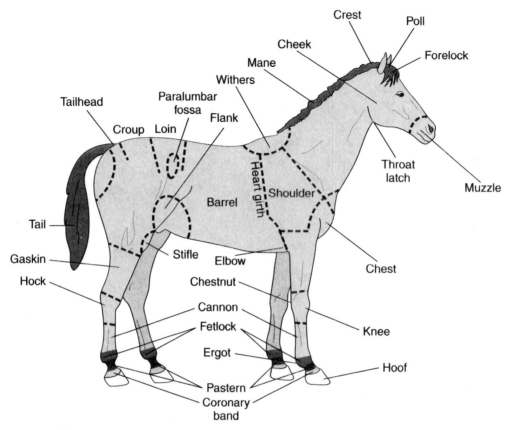

FIGURE 4–8 External anatomy of the horse

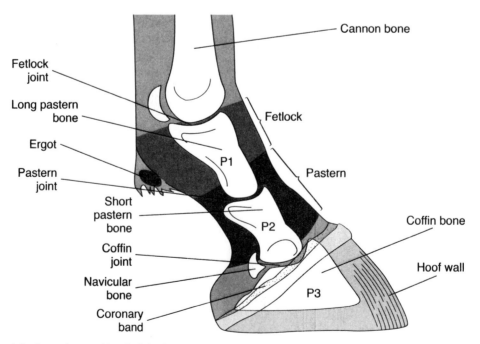

FIGURE 4–9 Parts of the lower leg and hoof of the horse

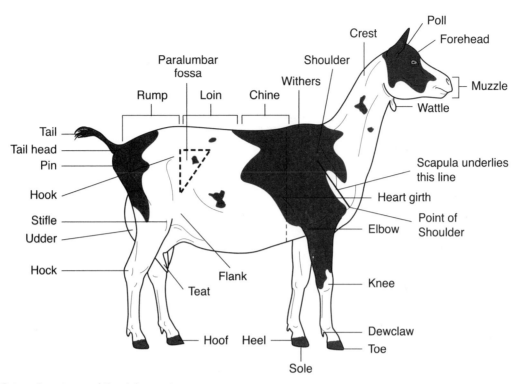

FIGURE 4–10 External anatomy of the dairy goat

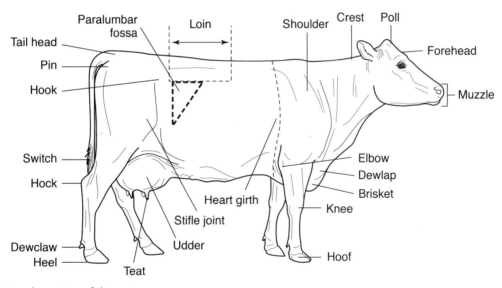

FIGURE 4–11 External anatomy of the cow

Dewlap   Loose skin under the throat and neck of an animal. Most often used in reference to cattle and rabbits.

Dock   The top portion of the tail. This term is also used for the surgical removal of all or part of the tail.

Fetlock (feht-lohk)   The joint formed between the cannon bone (third metacarpal) and the pastern (P1). The fetlock joint also contains two sesamoid bones.

Flank   The region where the barrel meets the hind leg.

Gaskin (gas-kihn)   The muscular region on the inside and the outside of the tibia in the hindquarter.

Heartgirth   The circumference of the area directly behind the withers and the elbows. Heartgirth measurement is related to weight, and a common way to estimate the weight of a horse is to use a weight tape to measure around the heartgirth.

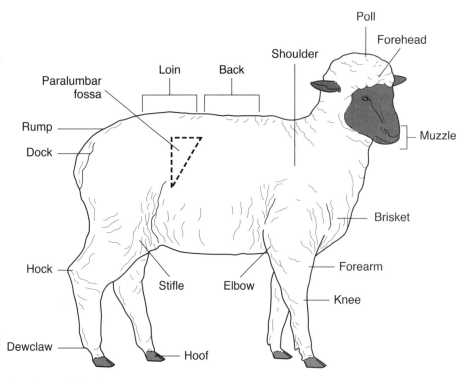

FIGURE 4–12 External anatomy of the sheep

Heel The back of the hoof.

Hoof (huhf) The hard foot of some animals. Animals with hooves are known as **ungulates.** Ungulates can have a single hoof, such as horses, or a split hoof, such as cows and pigs. Hooves are made of the same type of tissue as human fingernails, and can be trimmed with no discomfort to the animal.

Muzzle (muh-zuhl) The nose area of an animal, including the nostrils and lips.

Paralumbar fossa (pahr-ah-luhm-bahr fohs-ah) The hollow between the ribs and the hips (see Figure 4-12).

Poll (pōl) The area on the top of the head, between the ears.

Tailhead The part of the tail where it connects to the body.

Teat The nipple of the mammary gland.

Throatlatch (thrōt-lach) The region of the underside of the neck where the head and neck connect. Also known as the **throat** or **throttle.**

Toe The front of the hoof in ungulates, or the individual digit in clawed animals or poultry (see Figure 4-13).

Topline The term to describe the unit created by the back, loin, and croup or rump.

Udder The mammary gland of the female of the species.

Underline The bottom of the body from the elbows to the flank.

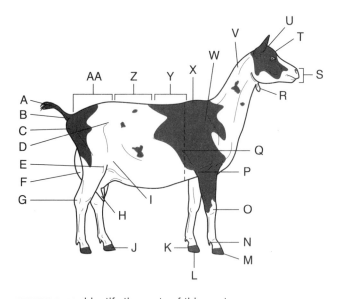

FIGURE 4–13 Identify the parts of this goat.

Parts of the Hoof. The hoof supports the weight of the entire animal, and is structured to do so with minimal problems (see Figure 4-14). The following are the part of a hoof:

Bars V-shaped structure on the bottom of the hoof leading from the heel toward the toe.

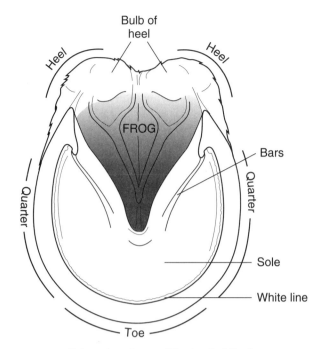

FIGURE 4–14 External anatomy of the hoof of the horse

Laminae (lahm-ihn-nah)   Structures inside the hoof. Laminae can be sensitive (with nerves) or insensitive (without nerves). The sensitive and insensitive laminae interlock in the hoof. Inflammation of the laminae is **laminitis** (lahm-ihn-ī-tihs), and is a serious illness in hooved animals.

Quarter   The region of the hoof on the sides of the hoof between the sole and the heels.

Sole   The bottom of the hoof.

White line   The area of the bottom of the hoof where the sole meets the hoof wall.

Species-Specific External Anatomy Terminology

The following are terms specific to each of the following species.

Bovine

Brisket (brihs-kiht)   The mass of muscle and tissue covering the chest.

Hooks   The protruding hip bones that are part of the pelvis closest to the head.

Pins   The protruding bones of the pelvis on either side of the tail.

Switch.   A tuft of hair at the end of the tail.

Frog   The arrow-shaped pad on the hoof of the horse. The frog is flexible and plays an important role in moving blood back up the leg.

Heel   The back part of the hoof. The bulb of the heel is the soft tissue on the heel, and is susceptible to injury.

Hoof wall   The hard outer layer of the hoof. The hoof wall primarily bears the weight of the hoof.

Porcine.   The following are terms specific to porcine (see Figure 4-15):

Ham   The heavily muscled upper thigh.

Jowl   The throat and cheek region.

Rump (ruhmp)   The region of the topline that is directly above the sacral vertebrae.

Snout (snowt)   The specific term for the upper lip and nose of swine.

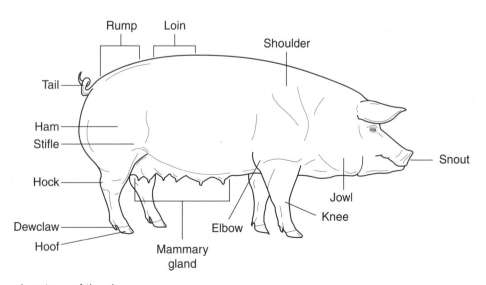

FIGURE 4–15 External anatomy of the pig

Caprine/Ovine

Brisket (brihs-kiht)   The mass of muscle and tissue covering the chest.

Chine (kīn)   The area from the withers to the loin. The thoracic vertebrae are under the chine region.

Horn butt   The region between the eyes and ears where **horns** have previously grown.

Wattle (what-tuhl)   Fleshy growths under the throat-latch of goats.

Equine

Chestnut (chĕs-nut)   Horny tissue on the inside of the legs of horses. Chestnuts are found above the knees in the front legs, and below the hocks in the hindquarters.

Ergot (ər-goht)   The horny mass, similar to a dew-claw, on the back of the fetlock. The ergot is surrounded by hair.

Forelock (fŏr-lohk)   The portion of the mane that grows from the poll down the forehead of the horse.

Frog   The triangular soft region of the hoof that rests between the bars on the bottom of the hoof. The frog is softer than the rest of the hoof, and assists in circulation in the leg.

Mane (mān)   The hair that grows from the top of the neck.

Splint bones   Another term for the first and fourth metacarpal or metatarsal bones that lie on the inside and outside of the cannon bone.

Canine/Feline. The following are terms specific to canines and felines (see Figure 4-16):

Dewclaw (doo-claw)   A remnant of an evolutionary toe that protrudes from the back of the metacarpals (metatarsals).

Pinna (pihn-ah)   The cartilage of the ear that is external to the head.

Rump   The region of the back between the hips and the tailhead.

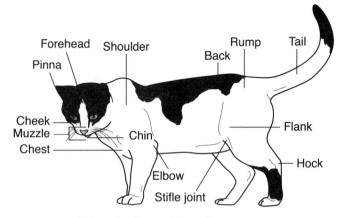

FIGURE 4–16 External antomy of the cat

Muscular Anatomy

The skeleton is the framework for the body, and muscles move that skeleton. Different types of muscles serve different roles, but they have some things in common. All muscles are made of groups of muscle cells called **muscle fibers** and expand and contract to create a specific action. Voluntary muscles are those that expand and contract in response to a conscious decision of the animal to make a movement; involuntary muscles are those that expand and contract without conscious thought (see Figure 4-17). Generally, **voluntary muscles** are involved in movement of the body, and **involuntary muscles** are involved in the functions necessary for life, such as the beating of the heart and breathing. The following is a list of the types of muscles found in animals:

Cardiac muscle (kar-dē-ahck)   Also known as the **striated** (strī-ā-tehd) **involuntary muscle.** The striated muscle is characterized by microscopic dark bands across the muscle fiber. The cardiac muscle is involuntary, and is started by a bundle of nerves in the heart. The **autonomic** (aw-tō-nah-mihk) nervous system, which controls involuntary muscles, can increase or decrease the heart rate, but does not initially stimulate the contraction of the muscle.

FIGURE 4–17 Different muscle types- a-skeletal, b-smooth, c-striated

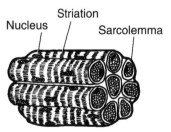

A) Skeletal

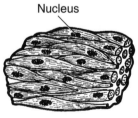

B) Smooth

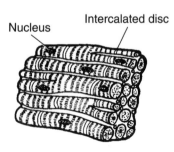

C) Cardiac

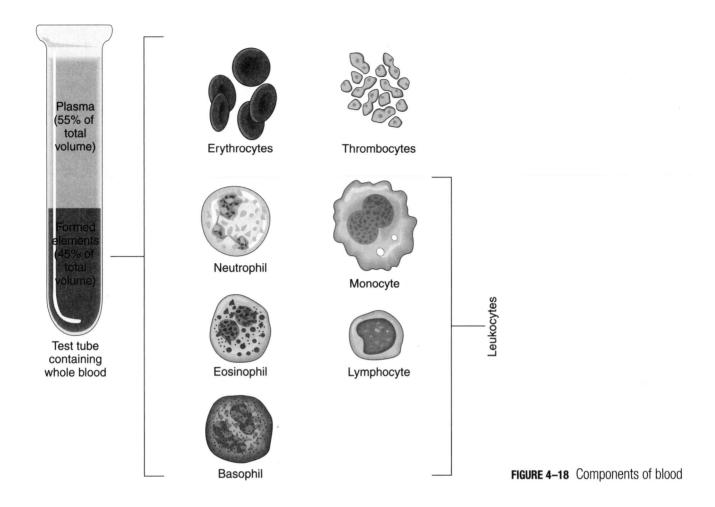

Plasma (55% of total volume)

Formed elements (45% of total volume)

Test tube containing whole blood

Erythrocytes

Thrombocytes

Neutrophil

Monocyte

Eosinophil

Lymphocyte

Basophil

Leukocytes

FIGURE 4–18 Components of blood

Skeletal muscle Skeletal muscles work to move the bones, and make animals **ambulatory** (ahm-bū-lah-tōr-ē), or able to move. Skeletal muscles are also known as **striated voluntary muscles.** Skeletal muscles respond to the voluntary nervous system; a message from the nerves to contract causes a muscle to contract, and removal of the signal causes the muscle to relax. Most skeletal muscles are found in pairs, to offset each other's action. **Abductor** (ab-duchk-tər) muscles move bone away from the body, and **adductor** (ahd-duhck-tər) muscles move bone toward the body. Skeletal muscle makes up the majority of the meat cuts that are used by consumers.

Smooth muscle Smooth muscles are also knows as **unstriated** (un-strī-ā-təd) **involuntary muscles.** These involuntary muscles are stimulated by the autonomic nervous system, which means they operate without conscious effort. Smooth muscles surround the internal organs and are not connected to the skeleton. These muscles move food through the digestive tract, and perform other basic bodily functions.

Circulatory System

The circulatory system moves blood throughout the system, carries nutrients to the cells, and brings waste products back to be released from the body. The circulatory system is made of the heart and blood vessels. The following terms refer to parts of the circulatory system:

Pulmonary (puhl-mah-nār-ē) The pulmonary circulatory system is the portion of the circulatory system that serves the lungs.

Systemic The systemic circulatory system is that which carries blood to the rest of the body.

Hepatic (heh-paht-ik) The hepatic circulatory system is a part of the systemic circulatory system, and carries blood from the internal organs to the liver, where the liver filters out waste. The filtered blood then returns to the systemic system.

Blood Vessels

Blood (bluhd) **vessels** Through vessels, blood transports nutrients, gasses, and waste throughout the body. Blood has the following components:

Plasma (plahz-mah)   The liquid portion of blood and about 60 percent of the total blood volume. Plasma contains water, hormones, and other proteins.

Platelets (plāt-lehts)   The smallest components in the blood, platelets are sticky particles that converge on sites of injury. Platelets stick together to create a clot to stop bleeding.

Red blood cells   Also known as **erythrocytes** (eh-rihth-rō-sītes), red blood cells carry oxygen and other gasses out to the body. In mammals, red blood cells do not have a nucleus; while birds do have nuclei in their red blood cells. Because mammalian red cells do not have a nucleus, they cannot generate energy, and have a short life span.

Thrombocyte (throhm-bō-sīt)   Another type of cell that causes clotting at the site of a wound to stop bleeding. Platelets do not have nuclei, and thrombocytes have nuclei.

White blood cells   Also known as **leukocytes** (loo-kō-sītes), white blood cells are fewer in number in the blood than red blood cells, and play an important role in the immune system. **Neutrophils** (nū-trō-fihls), **eosinophils** (ē-ō-sihn-ō-fihls), **basophils** (bā-sō-fihls), **monocytes** (mohn-ō-sīts), and **lymphocytes** (lihm-fō-sīts) are all types of white blood cells.

Arteries (ahr-tər-ēz)   Arteries carry blood away from the heart. Arteries have three layers, an outer layer of connective tissue, a middle layer of smooth muscle tissue, and an inner layer of **endothelial** (en-dō-thē-l-ē-əhl) tissue. The artery is flexible enough to absorb some of the pressure of the pumping of blood. The pulse of the heart pumping can be felt in arteries near the skin surface. The following are types of arteries:

Aorta (ā-ōr-tah)   The largest artery, the aorta leads from the heart and branches into smaller arteries to carry blood throughout the body.

Arterioles (ahr-tēr-ē-ōlz)   The smallest of arteries, arterioles are almost completely smooth muscle cell.

Pulmonary   Carries blood from the heart to the lungs. The pulmonary artery has two branches, one to each lung.

Capillaries (kahp-ih-lār-ēz)   Capillaries are the smallest of the blood vessels, and the walls are made up of cells that allow oxygen, nutrients, and waste products to be exchanged between the capillaries and the cells. Capillaries transition between the arterial system and the **venous** (vē-nuhs) system.

Vein (vēn)   Veins carry blood from the cells back to the heart and lungs. Whereas arteries carry nutrients and oxygen to the cells, veins carry waste products, such as carbon dioxide and water, back through the system for disposal. Veins are much less muscular than arteries, and have valves that stop the blood from flowing back in the veins. The following are types of veins:

Pulmonary   Two veins from each lung that carry oxygenated blood back to the heart for circulation throughout the rest of the body.

Vena cava (vē-nəh cā-vəh)   The large veins leading into the heart. The anterior or cranial vena cava carries blood from the head and upper body, and the posterior or caudal vena cava carries blood from the lower or hind part of the body.

Venules (vēn-yuhlz)   Small veins that lead from the capillaries to the larger veins.

Heart

The heart powers the circulatory system. The rhythmic pumping of the heart pushes blood through blood vessels throughout the body (see Figure 4-19). The following are parts of the heart:

Atrium (ā-trē-uhm)   The **atria** (ā-trē-ah) are two of the four chambers of the heart. Veins bring blood from the body into the atria. Blood enters the right atrium from throughout the body, and blood enters the left atrium from the lungs.

Coronary blood vessels (kōr-oh-nār-ē)   The blood vessels that serve the heart. Although blood from throughout the body travels through the heart, the coronary veins and arteries provide the nutrients and remove the waste from the heart muscle.

Ventricles (vehn-trih-kuhlz)   The ventricles do the pumping, and are more heavily muscled than the atria. Blood moves from the right atrium to the right ventricle. The right ventricle pumps blood to the lungs. The blood then comes from the lungs through the left atrium to the left ventricle. The left ventricle then pumps blood throughout the body.

Digestive System

The **digestive** (dī-jehs-təv) system takes feedstuffs and breaks them down so that the body can use the nutrients and energy from the food. Details on that process will be covered in Chapter 5 on nutrition. In this section, we will focus on the organs that make up the digestive system, which is also known as the **gastrointestinal** (gahs-trō-ihn-tehst-i-nəhl) tract (see Figure 4-20). Digestive tracts have evolved differently for different species, depending on their primary food source. The following are components of the digestive system in the order in which food would pass through the system:

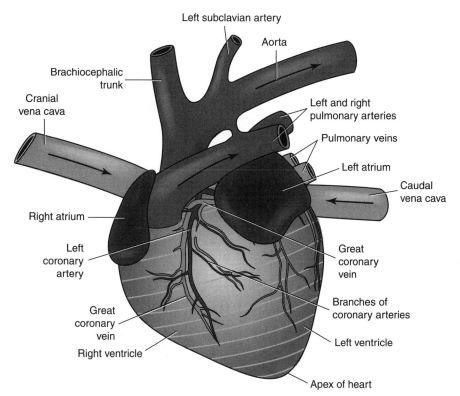

FIGURE 4–19 External anatomy of the canine heart

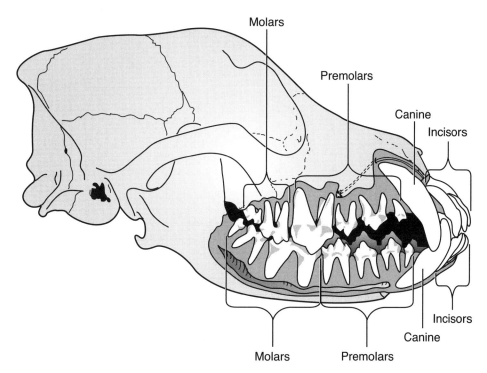

FIGURE 4–20 Dentition of an adult dog

Mouth   The mouth is the entrance to the digestive tract and contains many of the structures that begin the mechanical and chemical process of breaking down food.

Tongue (tuhng)   The muscular organ in the mouth that moves food during chewing, and has the taste buds, or **papillae** (pah-pihl-ā), that taste the food.

Teeth   The teeth begin the mechanical breakdown of food. Different species have different combinations of teeth types, which are divided into the following four primary types with different functions (see Figure 4-21).

Incisors (ihn-sī z-ōr)   The front teeth that cut through food.

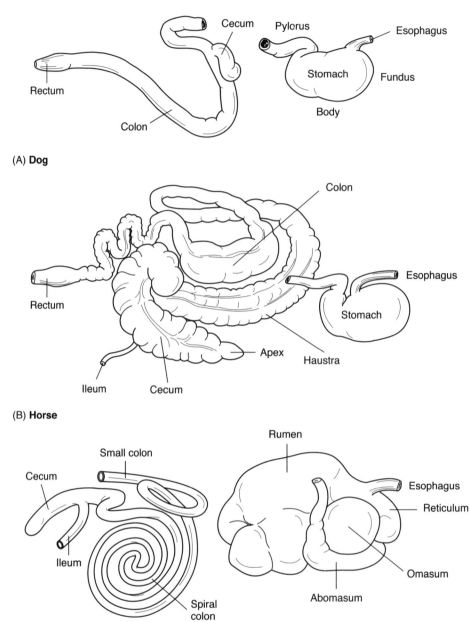

FIGURE 4–21 Digestive tracts (small intestinal segments eliminated for clarity): a-dog, b-horse, c-ruminant

Canines (kā-nōn)   The pointed teeth next to the incisors. Canines, or fangs, are more pronounced in predatory animals, and are used for grasping food.

Premolars (prē-mō-lahr)   The teeth immediately preceding the molars. Premolars are also sometimes called **bicuspids** (bī-cuhs-pĭd) because they have two points on the surface.

Molars (mō-lahr)   The large flat surfaced teeth in the back of the jaw that do most of the grinding of food.

Salivary glands (sahl-ih-vahr-ē glahndz)   The glands around the mouth that secrete **saliva** (sah-lī-vah), which introduces moisture and digestive enzymes to the food.

Esophagus (ē-sohf-ah-guhs)   The tube that carries food from the mouth to the stomach. The esophagus is a muscular tube that ends in a muscular ring, the **sphincter** (sfingk-tər), at the stomach.

Stomach   The stomach is a muscular organ that is the site of the majority of digestion. The stomach adds digestive chemicals to the food, and muscular contractions of the stomach assist in mechanical breakdown of food. Stomachs are divided into two types, **monogastric** (mon-nō-gahs-trihck) and **ruminant** (roo-mihn-ehnt).

Monogastric   The monogastric has a single stomach. Examples of monogastric animals are humans, dogs, cats, and pigs. In monogastric digestive systems, the majority of digestion occurs in the stomach. In the stomach, gastric juices (which are very acidic) are introduced to the food and break it down to prepare it for absorption. Feedstuffs are broken down into a liquid called **chyme** (kīm), which is then moved into the small intestine.

Ruminant The ruminant stomach has multiple compartments (see Figure 4-22). Examples of ruminant animals are cattle, sheep, goats, and llamas. The llama is a modified ruminant, and has three stomach compartments, instead of the four found in other ruminants. The following are the four compartments of the ruminant digestive system:

a

c

b

d

FIGURE 4–22 The linings of the 4 stomachs of a cow: a-rumen, b-reticulum, c-omasum, d-abomasum. The lining of the abomasum is very similar to the lining in the single stomach of monogastric animals.

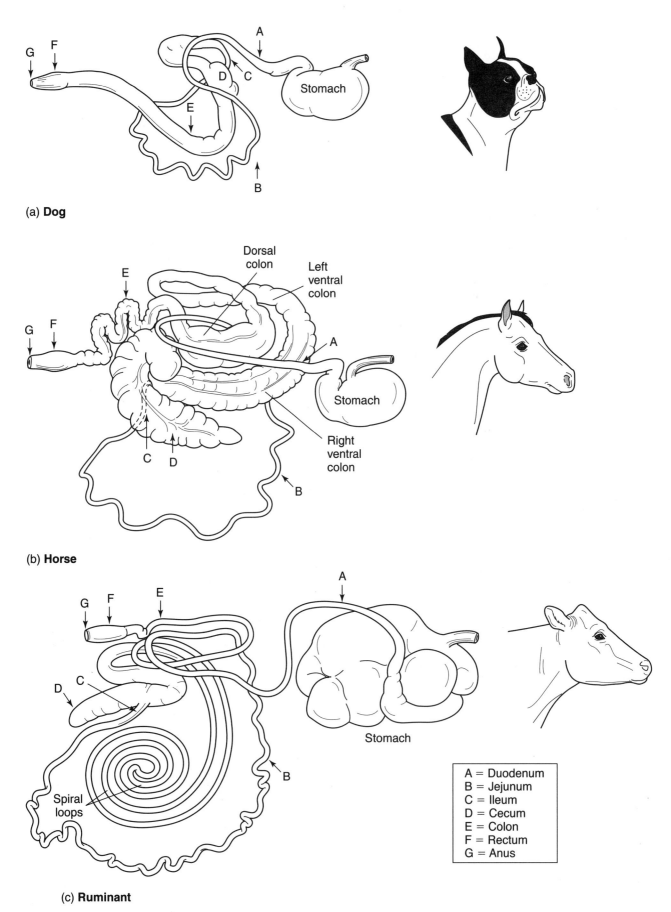

(a) **Dog**

(b) **Horse**

(c) **Ruminant**

A = Duodenum
B = Jejunum
C = Ileum
D = Cecum
E = Colon
F = Rectum
G = Anus

FIGURE 4–23 The intestinal tract of the dog (a), the horse (b), and the cow (c).

Rumen (roo-mehn)   The largest portion of the stomach of ruminant animals. The rumen is the site of fermentation of the foodstuff after ingestion. When animals have consumed enough food to fill the rumen, they lie down and **ruminate** (roo-mih-nāt), which is the regurgitation of foodstuff up the esophagus for further chewing.

Reticulum (re-tihck-yoo-luhm)   The portion of the stomach that contains a series of membranes that create a honeycomb-like effect. After initial consumption of food, the solid portions stay in the rumen, and the liquid portions move to the reticulum.

Omasum (ō-mā-suhm)   The muscular portion of the stomach that does some grinding, and squeezes water out of the food bolus.

Abomasum (ahb-ō-mā-suhm)   The "true" stomach. Activity in the abomasums of the ruminant is similar to the stomach in the nonruminant.

Small intestine   After initial digestion in the stomach, digestion continues in the small intestine. The small intestine is the site of the absorption of most nutrients. The small intestine includes the following three sections:

Duodenum (doo-wahd-nuhm or doo-ō-dē-nuhm)   The segment of the small intestine nearest the stomach.

Jejunum (jā-joo-nuhm)   The middle segment of the small intestine.

Ileum (ihl-ē-uhm)   The last segment of the small intestine.

Cecum (sē-kuhm)   The cecum is located where the small intestine meets the large intestine. In most species, the cecum does not have a major role in digestion; however, in horses, the cecum is a major site of digestion and fermentation of high-fiber feed.

Large intestine   The majority of nutrients are absorbed in the small intestine, and the large intestine absorbs water from what remains of the feedstuff (see Figure 4-23). The following are the parts of the large intestine:

Colon   The part of the large intestine from the cecum to the rectum.

Rectum (rehck-tuhm)   The end of the large intestine.

Anus (ā-nuhs)   The terminal end of the digestive tract. Feedstuffs that are left after digestion are excreted through the anus as **feces** (fē-sēz).

In addition to the digestive tract, which runs as a continuous tube through the body, the following organs play important roles in digestion and utilization of feedstuffs:

Liver (lĭv-əhr)   One of the largest glands in the body, the liver produces **bile** (bīl), which helps break down fat in foods. The **gall bladder** (gahl blahd-dər) is located in the liver, and stores bile in all animals except horses and rats, which continually excrete bile into the small intestine, instead of storing it in the gall bladder.

Pancreas (pahn-krē-uhs)   The pancreas produces the digestive enzymes **trypsin** (trihp-sihn), **lipase** (lī-pās), and **amylase** (ahm-ih-lās). The pancreas also produces the hormone **insulin** (ihn-suhl-ihn).

Avian Digestive System

The avian digestive system has some specific differences when compared to a mammalian digestive system (see Figure 4-24). The mouth in avian species does not contain teeth, and is simply used to pick up and swallow food. The esophagus leads to the **crop**, which stores the feed and adds saliva for moisture. From the crop, food moves through the **proventriculus** (prō-věn-trĭck-yoo-luhs), the glandular stomach, to the muscular stomach, also known as the **gizzard** (giz-erd). The muscular gizzard contracts to grind the feed into smaller pieces before it moves to the **small intestine** for nutrient absorption, the **large intestine** for absorption of water, and then remaining material is excreted through the **cloaca** (klō-ā-kah) and **vent** (vehnt). Avian species have **ceca** (sē-kah), the plural of cecum, but the exact function of the ceca is unclear.

Reproductive System

The **reproductive** (rē-prō-duhck-tihv) system provides for the production of the next generation of offspring. The study and science of reproduction is also known as **theriogenology** (thēr-ē-ō-jehn-ohl-ōj-ē). Reproductive organs are called **genitals** (jehn-ih-tahlz), and differ in males and females. In addition to producing **gametes** (gahm-ēts), which are the cells containing the genetic material from each parent, the genitals produce many of the chemicals called

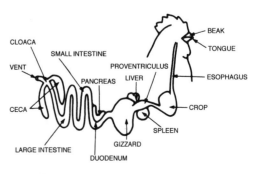

FIGURE 4–24 The digestive tract of the chicken.

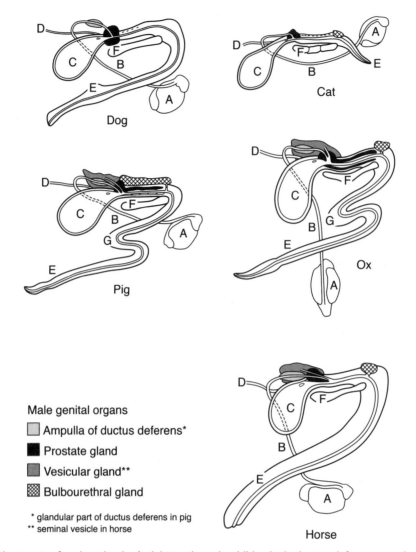

FIGURE 4–25 Reproductive tracts of male animals. A-right testis and epididymis, b-ductus deferens, c-urinary bladder, d-ureter, e-penis and extrapelvic urethra, f-pelvic sympysis, g-sigmoid flexure.

hormones (hōr-mōnz), which are involved in controlling the reproductive process.

Male Reproductive System

Male reproductive systems vary in appearance between species, but many of the anatomical parts serve similar purposes (see Figure 4-25). The following are parts of the male reproductive system:

Accessory glands   Glands such as the **prostate** (proh-stāt), **bulbourethral** (buhl-bō-yoo-rē-thrahl), and **vesicular** (vehs-ih-koo-lar) glands that produce fluids to nourish and protect sperm. Fluids produced by the accessory glands also flush urine from the urethra. There is some variation across species regarding accessory glands. Swine have an **ampulla** (ahmp-yoo-lah), and in horses, the vesicular glands

are known as **seminal vesicles** (sehm-ih-nahl vehs-ih-kuhlz).

Acrosome (ahk-rō-zōm)   A structure on the sperm cell containing enzymes that allow the sperm to penetrate the ovum for fertilization.

Cloaca (klō-ā-kah)   An external reproductive organ in male and female birds. In males, the cloaca holds sperm cells until the rudimentary penis transfers them to the female cloaca.

Epididymis (ehp-ih-dihd-ih-mihs)   A tube on the testicle that stores semen prior to ejaculation.

Leydig's cells (lih-dihgz sellz)   Cells that produce hormones such as testosterone.

Penis (pē-nihs)   The male sex organ that delivers sperm into the female reproductive tract. The penis also serves as the organ to facilitate excretion of urine from the body.

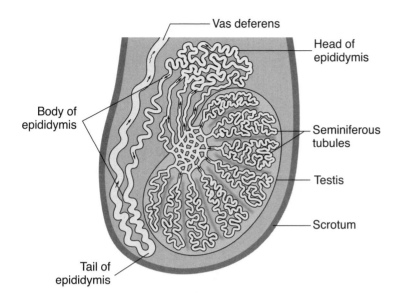

FIGURE 4–26 Cross section of the scrotum and associated structures

Scrotum (skrō-tuhm) The skin sac that encloses and supports the testes outside the male body (see Figure 4-26). The testicles of mammals must be cooler than the body temperature to allow production of live sperm.

Sigmoid flexure (sihg-moyd flehck-shər) An S-shaped bend in the penis of boars and bulls.

Sperm (spərm) Male sex cells (see Figure 4-27). The sperm carry the DNA that the male will contribute to the offspring. The sperm cell is divided into three major parts: the **head,** which contains the genetic information; the **body,** which connects the head to the tail; and the **tail,** which allows the sperm cell to swim.

Spermatazoa (spər-mah-tō-zō-ah) Another term for male sex cells. Singular form **spermatozoon** (spər-mah-tō-zō-uhn).

Testes (tehs-tēz) Glands that produce spermatozoa and male sex hormones. Singular form **testis** (tehs-tihs).

Testicles (tehs-tih-kuhlz) Another term for testes.

Urethra (yoo-rē-thrah) A tube that runs through the penis that carries urine from the urinary bladder, or semen from the testicles.

Vas deferens (vaz dehf-ər-ehnz) A tube that carries sperm from the epididymis through the penis.

Female Anatomy and Physiology

As in the male, the appearance of some aspects of the reproductive tract may have differences, but the functions are very similar (see Figure 4-28). The following are terms related to female anatomy:

Albumen (al-bū-mən) The protein-rich white of the egg. It is added to the **yolk** after **ovulation** when the

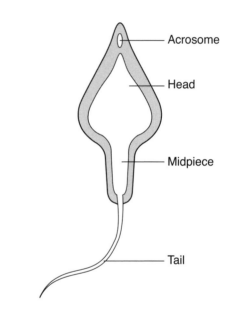

FIGURE 4–27 Anatomy of the sperm cell

yolk travels through the **magnum** portion of the **oviduct.**

Cervix (sihr-vihckz) The cervix separates the uterus from the more external reproductive organs. The cervix relaxes during **estrus** to allow the sperm to enter the reproductive tract. It then closes until the next estrus, or until it is time to deliver an offspring.

Corpus luteum (kōr-puhs loo-tē-uhm) Cells formed from the follicular cells after ovulation. The corpus luteum is yellow, and produces progesterone. Each ovulated follicle produces a corpus luteum, so

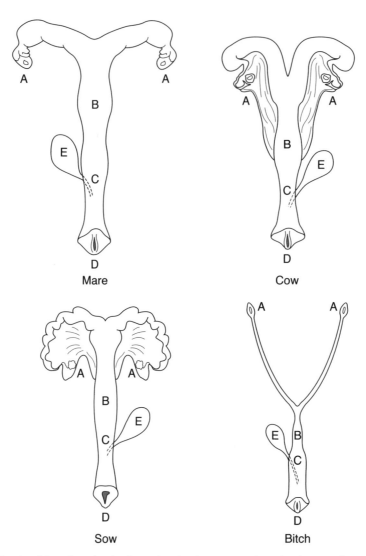

FIGURE 4–28 Reproductive tracts of female animals. A-ovaries, b-uterus, c-vagina, d-vulva, e-urinary bladder

animals that ovulate several follicles will have several **corpora lutea** (kōr-par-ah loo-tē-ah).

Fallopian tube (fah-lō-pē-ahn) Another term for oviduct. *Oviduct* is the preferred term in a professional atmosphere.

Follicle (fohl-lihck-kuhl) The sac on the ovary that contains the ova. As the follicle matures, it produces estrogen and enlarges until it ovulates and releases the ova. The following are types of follicles:

▶ **Primordial** Immature follicles in the ovary.
▶ **Secondary** Follicles that have begun to mature and produce estrogen.
▶ **Graafian** (grahf-ē-ahn fohl-lihck-kuhlz) Fully mature follicles that either ovulate or go through **atresia,** which means they stop development and are reabsorbed by the ovary.

Germ cell In avian species, the portion of the yolk that contains the genetic information.

Infundibulum (ihn-fuhn-dihb-yoo-luhm) An opening at the end of the oviduct near the ovary. The infundibulum, and the **fimbriae** (fihm-brē-ah) at the ends of the infundibulum, act as a "catcher's mitt" and catch the ova when they leave the ovary. The infundibulum and fimbriae are not connected to the ovary. In avian species, sperm are stored in the infundibulum until the egg (yolk) arrives for fertilization.

Mammary glands (mahm-mah-rē) Mammary glands are present in all mammals, and produce milk for the young (see Figure 4-29). The number of glands present varies with species. The following are components of mammary glands:

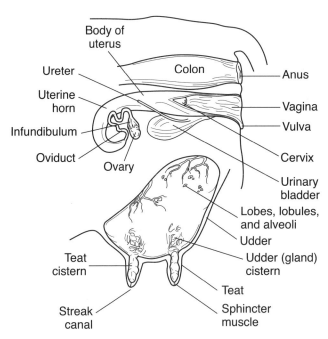

FIGURE 4–29 The mammary gland and reproductive tract of the cow

Alveoli (ahl-vē-ō-lī) The milk-producing part of the mammary gland. The saclike alveoli are arranged in lobes and lobules in the mammary gland.

Cisterns Areas of the mammary gland that store milk. The teat cistern is located in the teat, and the udder cistern is located in the udder.

Sphincter muscle A muscle that closes the **streak canal.**

Streak canal The opening from the teat cistern, through the teat, that milk travels through to leave the mammary gland.

Teat (tēt) The external portion of mammary gland that delivers milk to offspring.

Oocyte (ō-ō-sīt) An egg cell.

Ova (ō-vah) Egg cells. Singular form **ovum** (ō-vuhm).

Ovary (ō-vah-rē) The female gonad that produces ova as well as numerous female reproductive hormones. Mammalian species have two active ovaries, whereas avian species have one functional ovary.

Oviducts (ō-vih-duhckts) Tubes that lead from the ovary to the uterus. Oviducts are not connected to the ovary. Oocytes travel through the oviduct, and in most species, fertilization occurs in the oviduct. The following are parts of oviducts that are important in avian reproduction:

▶ **Magnum** (mayg-nuhm) In birds, this is the part of the oviduct where the albumen is secreted to surround the yolk.

▶ **Isthmus** (ihs-muhs) The part of the oviduct where shell membranes are added.

Placenta (plah-sehn-tah) The organ that joins the fetal unit with the uterus. The placenta exchanges nutrients and waste products between the fetus and the mother, and is expelled at parturition.

Sperm nests In birds, sperm nests are locations in the oviduct where sperm can survive for several weeks. Eggs are fertilized when they pass over the sperm nests. Longevity of sperm varies by individual rooster and by species.

Urethra (yoo-rē-thrah) The tube running from the bladder to the floor of the vagina that carries urine out of the body.

Uterus (yoo-tər-uhs) The muscular organ of the female. In mammals, this organ is the location for the development of the offspring. In avian species, this organ is where the shell is added to the egg. The uterus has the following three primary tissue layers:

▶ **Perimetrium** (pehr-ih-mē-trē-uhm) The outer uterine layer.

▶ **Myometrium** (mī-ō-mē-trē-uhm) The middle, muscular layer that contracts to deliver the fetus.

▶ **Endometrium** (ehn-dō-mē-trē-uhm) The inner layer that is primarily responsible for supporting the fetus.

Uterine body The portion of the uterus that is furthest from the oviduct. The cervix marks the end of the uterine body.

Uterine horn (yoo-tər-ehn) The portion of the uterus between the uterine body and the oviduct. Some species have **bicornuate** (bī-kōrn-yoo-āt) uteruses, which have large, well-defined uterine horns. In other species, the uterine body is the dominant structure and the uterine horns are quite small.

Vagina (vah-jī-nah) The structure that leads from the internal reproductive tract out of the body. The vagina receives the penis during copulation and serves as the "birth canal" during parturition.

Vulva (vuhl-vah) The external reproductive organ in females. Is some species, physiological changes of the vulva occur when they are in estrus, assisting in detection of estrus.

Endocrine System

Many of the processes in the body are controlled by **hormones** (hōr-mōnz), which are chemicals manufactured and released in the body in one set of cells, but that have an action on another set of cells. The study of these hormones, and how they work is called **endocrinology** (endō-crə-nahl-ahgē). The same basic hormones control processes in all mammalian species,

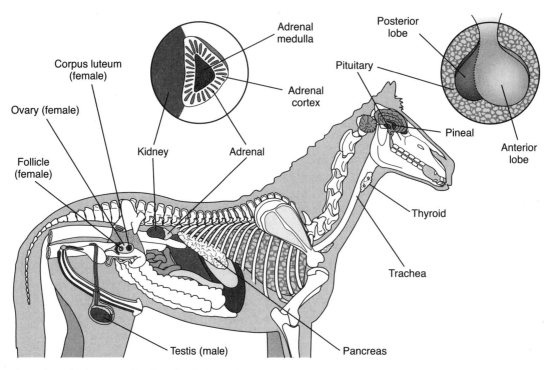

FIGURE 4–30 Location of primary endocrine glands in the horse

although species differ in timing and release patterns of the hormones. These differences are especially apparent in the control of the reproductive process (see Figure 4-30).

Endocrine Organs and Hormones They Produce

Most functions of the body are controlled by hormones. The following list identifies the important endocrine organs, and identifies the primary hormones produced by each of those organs:

Pineal gland (pī—nē-ahl) The pineal gland is located in the brain, and responds to changes in day length. The primary hormone produced by the pineal gland is **melatonin** (mehl-ah-tōn-ihn), which controls the timing of the reproductive season.

Hypothalamus (hī-pō-thahl-ah-muhs) Located at the base of the brain, the hypothalamus releases **gonadatropin-releasing hormone (GnRH)** (gə-nad-ah-trō-pən), which stimulates the anterior pituitary gland to produce follicle-stimulating hormones and luteinizing hormones.

Pituitary gland (pih-too-ih-tār-ē) The "master gland," the pituitary is divided into anterior and posterior sections and the hormones released from the pituitary gland have many roles (see Figure 4-31).

Anterior pituitary glands produce hormones that then act on target organs elsewhere in the body. The anterior pituitary controls many of the functions of

the body. The following hormones are all produced by the anterior pituitary gland, and then travel to other organs to elicit another response:

Adrenocorticotropic hormone (ACTH) (ahd-rēn-ō-kōr-tih-kō-trō-pihck) Acts on the adrenal glands to stimulate the release of **cortisol** and other **glucocorticoids.**

Follicle-stimulating hormone (FSH) A hormone released from the pituitary gland that stimulates the follicles to develop in the female, and stimulates sperm production in the male.

Growth hormone (GH) Stimulates growth of tissues in the body. Also known as **somatotropin.**

Luteinizing hormone (LH) (loo-tən-īz-ing) A hormone released from the anterior pituitary gland that has several roles in reproduction. LH stimulates ovulation. After ovulation occurs, LH acts on the ruptured follicle to create the corpus luteum. In males, LH stimulates the Leydig's cells to produce testosterone.

Prolactin (prō-lahck-tihn) Stimulates milk production.

Thyroid-stimulating hormone (TSH) Acts on the thyroid to stimulate production of thyroid hormones.

Posterior pituitary gland Secretes hormones that act directly on the target organs, whereas most of the hormones secreted from the anterior pituitary are "stimulating hormones." The following are hormones secreted by the posterior pituitary gland:

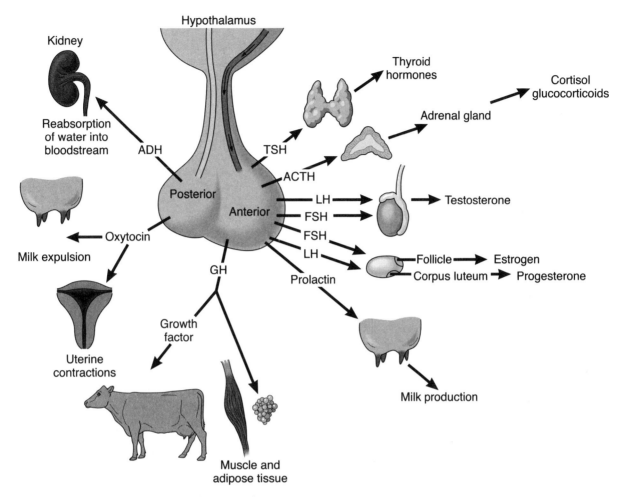

FIGURE 4–31 The pituitary glands and the hormones it secretes

Antidiuretic hormone (ADH) (ahn-tih-dī-yoo-reht-ihck) Also known as **vasopressin,** antidiuretic hormone controls the volume of water in the circulatory system through the kidneys.

Oxytocin (ahk-sē-tō-sən) Stimulates smooth muscle contraction in the mammary gland and the uterus, resulting in milk moving downward through the mammary gland (milk letdown) and uterine contractions.

Thyroid (thī-royd) A gland located in the neck, near the larynx. **Triiodothyronine (T3)** (trī-ī-ō-dō-thō-rō-nēn) and **thyroxine** (T4) (thī-rohcks-ihn) regulate metabolism.

Calcitonin (kahl-sih-tō-nihn) controls absorption of calcium.

Pancreas (pahn-krē-ahs) The pancreas secretes **insulin** (ihn-suh-lihn) and **glucagon** (gloo-kah-gohn), which are produced in specialized cells called the **Islets of Langerhans** (ī-lehts ohf lahng-ər-hahnz). Insulin lowers blood glucose by moving glucose into cells, and glucagon increases blood glucose by moving glucose out of cells into the blood.

Gonads In addition to production of gametes, gonads play an important endocrine role in the production of hormones. The following list indicates which hormones are produced by which gonads:

Testis

Testosterone (tehs-tahs-tər-ōn) The primary male reproductive hormone. Testosterone is produced by the Leydig's cells in the testes, and controls reproductive behavior and development of secondary sex characteristics in males.

Ovary

Estrogen (ehs-trə-gən) The dominant female reproductive hormone, estrogen is produced by the follicles on the ovary, controls reproductive behavior and development of secondary female characteristics, and acts upon the uterus to prepare it for the arrival of a zygote.

Progesterone (prō-jehs-tə-rōn) The female primary hormone during pregnancy. Progesterone is produced by corpora lutea and maintains pregnancy in the female.

Adrenal glands   Located near the kidneys, the adrenal glands produce hormones in the **adrenal cortex** or in the **adrenal medulla**. The adrenal cortex is the outer layer of the adrenal gland, and produces **glucocorticoids** (gloo-kō-kōr-tih-koydz) and **mineralocorticoids** (mihn-ər-ahl-ō-kōr-tih-koydz). Glucocorticoids control metabolism of fats, proteins, and carbohydrates, and have anti-inflammatory capabilities. Mineralocorticoids control water and mineral balance in the body. The adrenal medulla is the center of the adrenal gland and produces **epinephrine** (ehp-ih-nehf-rihn) and **norepinephrine** (nōr-ehp-ih-nehf-rihn). Epinephrine and norepinephrine are released in response to stress, and increase the heart rate, blood pressure, and the amount of blood glucose available for the muscles.

Prostaglandin (prahs-tah-glan-dən)   A broad category of hormones that are produced in various cell types. The most important one in animal science is prostaglandin F2-alpha, a hormone which breaks down the corpora lutea, which results in the halting of progesterone production, and the progression of the female through the estrous cycle. Prostaglandin can be produced in a wide variety of cell and tissue types, and is often used to manipulate the estrous cycle in mammals.

Respiratory System

The **respiratory** (rehs-pih-rah-tōr-ē) system brings oxygen into the body and removes carbon dioxide from the body (see Figure 4-32). **Respiration** (rehs-pər-ā-shuhn) is the exchange of gasses, and can occur either in the cells (internal or cellular respiration) or in the lungs (external respiration). In mammals, the **lungs** are the primary organs of the respiratory system. The **diaphragm** (dī-əh-frahm) is a large muscle below the lungs. Contraction and relaxation of the diaphragm expands and contracts the chest cavity to pull air in and force air out of the lungs.

Mammalian Respiratory System

The following are components of the respiratory system of mammals:

Nostrils (nahs-trihls)   The first part of the respiratory system is the nostrils. The nostrils are used to draw air into the respiratory system. Some animals can also inhale air through their mouths.

Nasal cavity (nā-zəhl)   The cavity just inside the nostrils. Air is warmed and moisture is added in the nasal cavity. This is also where particles of dust are removed, and the **olfactory receptors** (ohl-fahck-tōr-ē rē-sĕp-tōrz) detect smells.

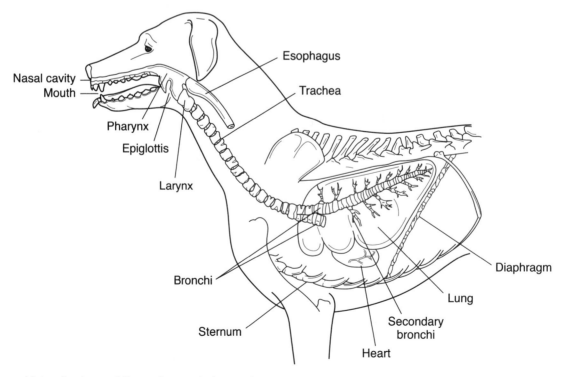

FIGURE 4–32 Major structures of the canine respiratory system

Pharynx (făr-ihnks)   The area in the back of the throat where the respiratory and digestive systems meet. In animals that breath through both the nose and mouth, air from both sources meets in the pharynx. The **epiglottis** (ehp-ih-gloht-his), which prevents food from entering the lungs, is located in the pharynx.

Larynx (lăr-ihnks)   The cartilage structure in the throat that contains the voice box and the vocal cords.

Trachea (trāk-ē-ah)   The trachea is also called the windpipe, and carries air to the bronchi. The trachea is made of rings of cartilage that help it maintain its shape. The trachea divides into two branches called **bronchi** (brohng-kī) in the chest cavity.

Bronchioles (brohng-kē-ōhlz)   Bronchi continue to branch in the lungs. The smaller branches of bronchi are bronchioles (see Figure 4-33).

Alveoli (ahl-vē-ō-lī)   The terminal ends of the bronchioles. Alveoli have extremely thin walls and interface directly with capillaries. The actual exchange of gasses occurs in the alveoli. Singular: **alveoulus** (ahl-vē-ō-luhs).

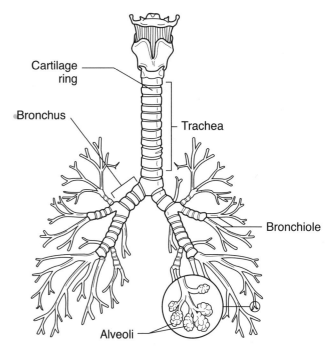

FIGURE 4–33 Lower respiratory tract structures

Avian Respiratory System

There are some significant differences between the avian and the mammalian respiratory systems (see Figure 4-34). Avian species have a **nasal chamber** that opens into the mouth, where the air then travels to the trachea. Instead of the pharynx and larynx that exist in

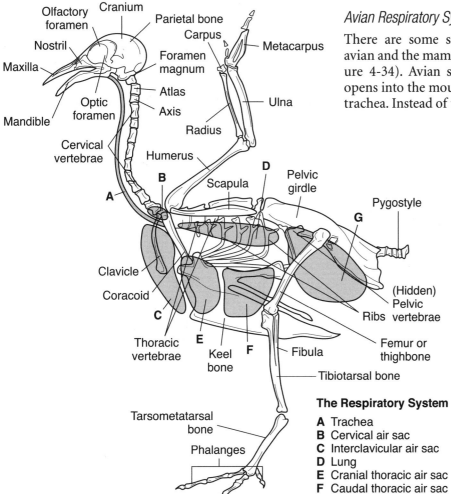

The Respiratory System

A Trachea
B Cervical air sac
C Interclavicular air sac
D Lung
E Cranial thoracic air sac
F Caudal thoracic air sac
G Abdominal air sac

FIGURE 4–34 Major structures of the avian skeleton and respiratory system

mammals, birds have a **syrinx** (sehr-ihncks) at the end of the trachea that produces sound. Birds also have **air sacs** at the ends of the bronchi and into the bones. Gas exchange occurs through **air capillaries,** instead of through alveoli. Contraction and relaxation of muscles in the chest and abdomen create breathing. Birds do not have a diaphragm.

Urinary System

The urinary system removes liquid waste from the body. Many of the parts of the urinary system are shared with the reproductive system (see Figure 4-35). The **kidney** is the primary organ of the urinary system, and filters waste products from the blood. In addition to filtering waste out of the system, the kidney plays an important role in maintaining proper hydration in the body. The kidneys remove excess water when it is present, and conserve water when it is scarce. The following are the parts of the kidney and urinary system:

Nephron (nehf-rohn)   The part of the kidney that filters the blood. The **glomeruli** (glō-mər-yoo-lī) are the capillaries that are part of the nephron.

Ureters (yoo-rē-tərz)   The paired tubes that carry urine from the kidney to the urinary bladder.

Urinary bladder   The organ where urine is stored until it is excreted from the system.

Urethra (yoo-rē-thrah)   The tube that carries urine from the urinary bladder outside the body. In males, the urethra also carries semen.

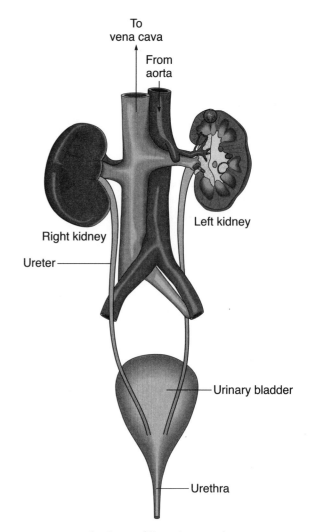

FIGURE 4–35 Anatomy of the urinary system

STUDY QUESTIONS

Match the correct skeletal part with the correct external part.

Skeletal Part	External part
1. _____ Humerus	a. Stifle
2. _____ Cranium	b. Coffin bone
3. _____ P3	c. Croup
4. _____ Third metacarpal	d. Arm
5. _____ Thoracic vertebra	e. Hock
6. _____ Carpal joint	f. Skull
7. _____ Tarsal joint	g. Cannon bone
8. _____ Patellar joint	h. Back
9. _____ Sacral vertebrae	i. Knee

10. Where in the digestive tract does the most absorption of nutrients take place?
    a. Stomach
    b. Small intestine
    c. Large intestine
    d. Cecum

11. Describe the structural differences between a vein and an artery.

12. What four parts make up the ruminant stomach, and what is the role of each part?

13. List the three muscle types, and give an example of where in the body each muscle type is found.

14. List the types of connective tissue, and give an example of where each type of connective tissue would be found.

15. What are the parts of the pulmonary circulatory system?

16. What types of feathers are found on the wings of birds?

17. Properly label the parts of the heart indicated in Figure 4-36.

18. Label the parts of the sperm cell, and identify the primary role of each part.

19. List the parts of the female mammalian reproductive tract in order, from the ovary to the vulva.

20. List the bones in the front leg of the cow in order from the ground to the top.

21. Name two parts of the avian female reproductive tract that are not found in the mammalian reproductive tract.

22. What is the purpose of "sperm nests" and in what species do they exist?

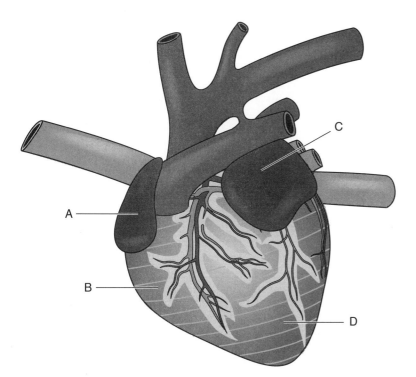

FIGURE 4–36 Label the parts of the heart indicated

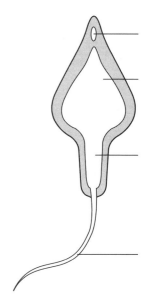

FIGURE 4–37 Label the parts of the sperm cell

Chapter 5
Animal Nutrition

Nutrients (noo-trē-əhnts) are the compounds in food that support growth, development, and maintenance of the body. **Nutrition** (noo-trish-shuhn), the study of these nutrients and how the body uses them, is one of the foundations of animal management, and a primary area of focus in animal sciences. Arguably, nutrition has the greatest potential impact on an animal being able to perform the purpose for which it has been bred. The genetics of the animal give it the potential to perform to a certain level, but that potential will not be fulfilled if the animal's nutritional needs are not met. Nutrition is a complex field. It is important to understand not only the nutrients, but also the foods that provide those nutrients, and the processes in the body that change the raw foodstuffs into a form that is usable by the body. Many of the nutrients have an optimal level for feeding. Either too much or too little can have negative effects on the health of an animal. Many diseases that challenge animal scientists are nutritionally related.

The study of animal nutrition also has direct human impacts. Many of the principles of nutrition that we have learned from doing research on animals have been applied to human nutrition. Currently, many of the nutrition-related issues facing humans, such as diabetes and obesity, are being researched using animal models.

Different types of animals gain their nutrients from different food sources. Their digestive systems have evolved to best utilize the nutrients from their food source of choice. As we seek to meet an animal's nutritional needs with commercial diets, it is important to understand how their digestive systems evolved, and the components of their original diets.

One part of the digestive tract that is significantly different between species is their dentition, or teeth (see Figure 5-1). Animals are often classified into the following groups based on the primary source of food:

Carnivores (kahr-nih-vōrs)   are animals that get most of their food from a meat source. Some species, like dogs, are preferential carnivores, whereas others, like cats, are **obligate carnivores.** Obligate carnivores are animals that must have meat in their diet to gain all of the nutrients they need. All carnivores are animals that we think of as predators, and have monogastric digestive systems.

Herbivores (hərb-ih-vōrs)   are animals that get most of their nutrients from plant sources. Some herbivores eat plant sources such as grass, and others are browsers, and eat plants that are woody in nature. Some herbivores are ruminants, and others are monogastrics. Examples of herbivores are horses, cattle, and sheep.

Omnivores (ohm-nē-vōrs)   are animals that eat a combination of plant- and meat-based food sources. Examples of omnivores are pigs, humans, and some birds.

Carnivore

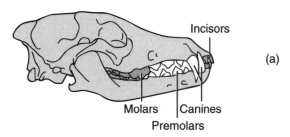

(a)

Herbivore

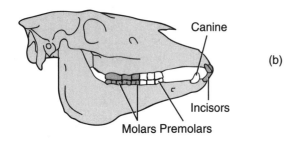

(b)

Omnivore

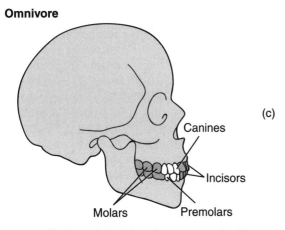

(c)

FIGURE 5–1 Skulls and dentition of a carnivore, herbivore, and ominvore. Note the omnivore has both tearing teeth, like the carnivore, and grinding teeth, like the omnivore.

NUTRIENTS

Nutrients can be divided into categories based on their chemical makeup and the purpose they serve for the body. Nutrients come from a variety of foodstuffs, and most foodstuffs contain several nutrients, although the amount of each nutrient in a particular food stuff is highly variable.

Carbohydrates

Carbohydrates (car-bō-hī-drāts)   Made up of carbon, oxygen, and hydrogen, carbohydrates are one of the primary sources of energy in the diet. Carbohydrates are burned for energy, and excess carbohydrates are stored in the body as fat.

Cellulose (sehl-yoo-lōs)   One of the most important carbohydrates in the diets of herbivores. Cellulose is found in most plant sources and is an important source of glucose that the body converts to energy.

Fiber   A type of carbohydrate classified as a complex carbohydrate. **Cellulose** (sell-yoo-lōs) and **lignin** (lihg-nəhn) are the specific carbohydrates that are in fiber-based feeds like grass and hay. The digestive tracts of ruminant animals can better utilize fiber than the digestive tracts of carnivores. Fiber is important in the diet to facilitate movement of feed through the digestive tract. In animals that are evolutionary herbivores, such as horses, it is important to maintain sufficient fiber in the diet to maintain the health of the digestive tract.

Sugars   A type of carbohydrate classified as a simple carbohydrate. Sugars are easily digested and found in grain-based feedstuffs.

Starches   A type of carbohydrate classified as a simple carbohydrate. Starches are easily digested to sugars, the other simple carbohydrate. As with sugars, starches are found in grain-based feedstuffs.

Fats

Fats are made up of carbon, oxygen, and hydrogen, and provide energy in the diet. Dietary fats can either be solid at body temperature (fats), or liquid at body temperature (oils). Fats are much more energy-dense that carbohydrates, with 2.25 times the calories per gram as carbohydrates. Fats can increase the palatability and energy density of feeds. However, excessive fat can lead to feed rancidity and make the feed unpalatable.

Minerals

Minerals (mihn-ehr-ahls) do not contain carbon, so they are inorganic. Minerals comprise a very small percentage of the diet, but are vital for normal growth and development. Minerals are involved in regulation of body systems, as well as proper growth of bones, teeth, and other tissue. Requirements for minerals vary, and care must be taken to meet the animals' mineral requirements without exceeding them. Deficiency and excess of minerals can both lead to poor health in animals. Minerals are divided into macro- and microminerals, and important minerals are listed on next page:

Macrominerals Minerals that are needed in relatively large amounts. The following is a list of macrominerals:

Calcium (kahl-sē-uhm) A mineral that is vital to the development of bones, muscle, and nerve activity.

Phosphorus (fahs-fər-uhs) A mineral that works with calcium in the skeleton, and helps with energy transfer throughout the body. It is important to keep phosphorus and calcium in proper balance in the diet.

Magnesium (māhg-nēz-ē-uhm) A mineral that, along with calcium and phosphorus, is involved with bone development.

Salt Sodium chloride is one of the most important minerals in the diet. Sodium and chloride are important in maintaining the proper water balance in the body. Salt is excreted through perspiration, and should always be available in the diet, or as a supplement.

Microminerals Minerals that are needed in relatively small amounts. Also knows as **trace minerals.** The following is a list of trace minerals:

Cobalt (kō-bahlt) Part of vitamin B12.

Copper (kahp-per) A mineral in red blood cells, skin, and hair pigment.

Iodine (ī-ah-dīn) A mineral that is essential in the **thyroid** (thī-royd) hormone, which controls metabolism.

Iron A component of red blood cells that helps carry oxygen in blood. Iron deficiency is known as **anemia** (ah-nēm-ē-ah).

Selenium (seh-lēn-ē-uhm) A mineral that is involved in muscle development and function.

Zinc A component of enzymes and an important mineral in normal skin and hair growth.

Proteins

Proteins (prō-tēn) are made up of **amino acids** (ah-mē-nō ahs-ihdz) that are derived from the diet. These amino acids are formed from carbon, hydrogen, oxygen, and nitrogen. Some amino acids contain other elements such as iron, phosphorus, and sulfur. Proteins are known as "building blocks" because of the vital role they play in the formation of everything from muscle to skin and hair and internal organs. Amino acids are classified as essential and nonessential. The following list defines which amino acids fit into each category:

Essential amino acids Essential amino acids must be provided in the diet because they cannot be metabolized in the body. There is some variation of essential amino acids with species. All animals require **methionine** (meh-thī-ō-nihn), **arginine** (ahr-jih-nēhn), **threonine** (thrē-ō-nihn), **tryptophan** (trihp-tō-fān), **valine** (vahl-ihn), **histidine** (hihs-tih-dihn), **isoleucine** (ī-sō-loo-sēn), **leucine** (loo-sēn), **lysine** (lī-sēn), and **phenylalanine** (fehn-ehl-ahl-ah-nēhn). Poultry also require **glycine** (glī-sēn) and **proline** (prō-lēn), and cats also require **taurine** (tawr-ēn).

Nonessential amino acids These amino acids do not need to be provided in the diet because they can be manufactured in the body from other amino acids. Different species can synthesize different amino acids.

Vitamins

Vitamins (vīt-ah-mihns) are organic compounds that are needed in small amounts in the diet. Vitamins are placed in the following two major classifications, fat-soluble and water-soluble.

Fat-soluble vitamins These vitamins can be dissolved in fat, and are stored in fat in the body. Care must be taken with fat-soluble vitamins to meet the needs of the animal without exceeding them. Because fat-soluble vitamins can be stored in the body, they can rise to toxic levels. The following are fat-soluble vitamins:

Vitamin A Involved in vision, skin health, and bone growth.

Vitamin D Involved in use and absorption of calcium, vital for proper bone growth and development.

Vitamin E Protects cells from damage and assists in muscle growth and immune function.

Vitamin K Essential in blood clotting.

Water-soluble vitamins These vitamins can be dissolved in water, and excesses are excreted through urine. Because the body can excrete water-soluble vitamins, toxicity is rarely seen. The following are water-soluble vitamins:

B vitamins The B vitamins include **thiamin** (thī-ah-mihn), **riboflavin** (rī-bō-flā-vəhn), **niacin** (nī-ah-sihn), **pyridoxine** (pēr-eh-docks-ēn), **pantothenic acid** (pahn-tō-thehn-ĭk), **biotin** (bī-ō-tehn), **folic acid** (fō-lihk), **benzoic acid** (behn-zō-ihk), **choline** (kō-lēn), and **vitamin B12.**

Vitamin C This water-soluble vitamin is available through many forages. Although most animals gain enough vitamin C through their diet, guinea pigs must receive vitamin C in the diet.

Water

Water is the single most important nutrient, and one that is often forgotten. Animals can live longer without food, or with inadequate food, than without water. Water is vital to allow the body to use other nutrients and for normal bodily function. All animals should have a constant supply of clean, fresh water. Animals may not consume water that is stale or dirty. Lack of water can lead to a range of disease states, such as colic in horses, and feline lower urinary tract disease in cats. Water need increases in hot weather, when animals are exercising, and during lactation. Dehydration (lack of water in the body) may also occur in cold weather, when the water supply is frozen, or the animals do not drink enough water because it is too cold.

PHYSIOLOGY OF DIGESTION

Ruminant and Nonruminant Digestion

Digestion is the process by which the body breaks down foodstuffs into the nutrients needed by the body. Digestion has chemical phases, which are performed by **enzymes,** and mechanical phases, where the foodstuffs are broken down by physical movement. After the digestive tract has broken down the foodstuffs, nutrients are absorbed into the blood stream and distributed throughout the body. In most species, this absorption takes place in the small intestine. Ruminant and nonruminant animals have different capacities to digest and use nutrients in feed, and varying capacities of feed they can consume.

Digestion begins in the mouth, with the chewing of food into a bolus (mechanical digestion), which is swallowed. In the mouth, saliva is added to the food bolus. The saliva contains the enzymes salivary **amylase** (ahm-ih-lās) and salivary **maltase** (mahl-tās), which begin to break down the foodstuff (chemical digestion). The process of chewing breaks the food into smaller pieces, which increases the surface area and provides more opportunity for the enzymes to begin chemical digestion.

Following chewing, the food bolus is swallowed, and travels down the **esophagus.** The esophagus uses **peristaltic** (pehr-ih-stahl-tihk), or wavelike, movement to move the bolus to the stomach.

In ruminant animals, the food bolus is swallowed before it is completely chewed. Solid food matter goes to the **rumen,** and liquid food matter goes to the **reticulum.** There is no physical barrier between the rumen and the reticulum, so food matter moves freely between the areas. After consuming a quantity of feed, the ruminant animal regurgitates solid food from the rumen back up to the mouth, and rechews the bolus. This rechewing is called **rumination,** or chewing cud. In the rumen and reticulum, bacteria and protozoa further break down the feed. The bacteria and protozoa also produce vitamins that the animal can then absorb and use. Furthermore, as the bacteria and protozoa die, they remain in the digestive tract and are digested as sources of protein.

Maintaining a healthy population of microorganisms in the rumen is vital to proper function of the digestive system. A diet that provides adequate forage for the microbes is necessary to maintain a healthy population. Feed of larger particle sizes maximizes the efficiency of the microbes in the rumen. The presence of these microbes allows the ruminant animal to efficiently use lower quality protein and forages than animals that do not have the capacity to bacterially ferment feed. From the rumen and reticulum, the food material (**digesta**) moves on to the abomasum, or true stomach, of the ruminant. From the abomasum on, digestion in the ruminant is similar to digestion in the nonruminant animal. In young ruminant animals that are consuming milk, the **reticular groove** can contract, making a route for the milk to bypass the immature rumen, directly to the abomasum, which digests the milk.

In the stomach, the stomach muscles contract on the digesta to continue to break it down, and to mix in the hydrochloric acid and enzymes secreted by the stomach. **Pepsin** (pehp-sihn) is the enzyme that breaks down proteins, and **lipase** (lī-pās) breaks down the fats. As the stomach contracts, the semiliquid parts are moved out of the stomach into the small intestine. This partially digested mass is called **chyme** (kīm).

The small intestine is the site of the final stages of digestion and absorption of nutrients. The enzymes **trypsin** (trihp-sihn), **pancreatic amylase,** and **pancreatic lipase** are added to the chyme. The trypsin continues to break down proteins, the amylase breaks down starches, and the lipase breaks down any remaining fats. Bile is secreted from the gall bladder into the small intestine where it assists the lipase in breaking down fats. Horses do not have a gall bladder, so bile is secreted from the liver. The small intestine is lined with small fingerlike projections called **villi** (vihl-ī), which are the site of nutrient absorption. The presence of villi increases the surface area of the small intestine, allowing it to absorb more nutrients.

The **cecum** (sē-kuhm) is located where the small intestine meets the large intestine (see Figure 5-2). In most species, the cecum has no clear role; however, horses have the capacity to bacterially digest forages in the cecum. Although less efficient in regard to maximizing the value of the roughage than the rumen, the cecum allows the horse to benefit from a primarily roughage diet. In fact, a high-roughage diet is important to maintaining the gastrointestinal (GI) tract health of the horse.

The **large intestine** is the final part of the GI tract. The primary role of the large intestine is to absorb the water remaining after the digestion of the foodstuff. Because horses digest feed in the cecum after it has been through the small intestine, nutrients resulting from

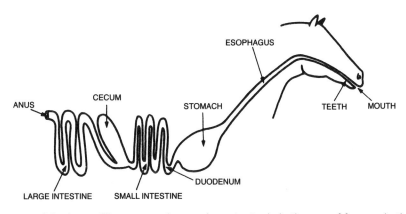

FIGURE 5–2 The digestive tract of the horse. The cecum plays an important role in the use of forages in the diet of the horse.

TABLE 5–1
Capacity of the digestive tract in livestock species

Organ	Swine (qts)	Swine (liters)	Equine (qts)	Equine (liters)	Bovine (qts)	Bovine (liters)	Ovine/Caprine (qts)	Ovine/Caprine (liters)
Rumen					80–192	75.7–181.6	25	23.6
Reticulum					4–12	3.8–11.4	2	1.9
Omasum					8–20	7.6–18.9	1	0.9
Abomasum					8–24	7.6–22.7	4	3.8
Nonruminant stomach	8	7.57	8–19	7.6–18				
Small intestine	10	9.5	27–67	25.5–63.4	65–69	61.5–65.3	10	9.5
Cecum	1–1.5	0.95–1.4	14–35	13.2–33.1	10	9.5	1	0.9
Large intestine	9–11	8.5–10.4	41–100	38.8–94.6	25–40	23.6–37.8	5–6	4.7–5.7
Total	28–30.5	26.5–28.87	90–221	85.1–209.1	200–367	189.3–347.2	48–49	45.2–46.2

Ranges indicate different ages, breeds, and sizes.

that digestion are absorbed in the large intestine. Unused feed material is moved through the large intestine to the **rectum** (rehck-tuhm), the distal portion of the large intestine. It is then excreted from the large intestine as feces through the **anus** (ā-nuhs), which is the exterior opening at the end of the digestive tract. Significant feed matter in the feces indicates that the animal is not digesting the food properly. In these cases, steps should be taken to determine the source of the problem.

Digestion in Poultry

The digestive tract of poultry has some different aspects when compared to mammals. Poultry do not have teeth, so chewing and addition of saliva are not part of the digestive process. Feed is consumed, swallowed, and stored in the **crop**. The crop is a thin-walled organ and is not a site of digestion. From the crop, food goes through the **glandular stomach** where enzymes and hydrochloric acid are added. Food then moves to the **gizzard,** a heavily muscled organ that contracts and grinds the food into smaller particles. Sometimes **grit,** a stone-type product, is added to the diet of birds to facilitate this grinding process in the gizzard. After leaving the gizzard, digestion is similar to that in the previously discussed species. Additional enzymes are added in the small intestine, which is the site of nutrient absorption. Feed material that is not digested is excreted as feces.

FEED ANALYSIS

Feedstuffs must be analyzed to determine what nutrients they can provide. Although certain feedstuff will have average amounts of nutrients, the amount of nutrients in any particular feedstuff can vary. For example, forages that are harvested at different points in maturation will contain different nutrients. Storage and handling can also impact the nutrients available in feed. Understanding the components of feed analysis is also important in evaluating commercial feeds that are available for livestock and companion animals. **Proximate analysis** (see Figure 5-3) is the chemical analysis of foodstuffs. Proximate analysis quantifies the water, ash, crude protein, ether extract, crude fiber, and nitrogen-free extract in a feedstuff. The following list defines each of these components of proximate analysis:

Ash   The mineral content of feed.

Crude fiber   Cellulose, hemicelluclose, and insoluble lignin in the feed.

Crude protein   The total amount of nitrogen in a feed. This includes the nitrogen that is in the form of protein, and other nitrogen that is not incorporated into protein. Care must be used when evaluating the usefulness of crude protein in monogastric animals. Although ruminants can utilize most forms of nitrogen, monogastric animals need amino acids to provide for their protein requirement.

Dry matter (DM)   The portion of the feed that is not water. The percentages of nutrients in a feed are based on dry matter, which gives the most accurate measurement of the amount of the nutrient the animal is receiving. Percentages may also be reflected in an **as-fed** basis. It is especially important to understand dry matter and "as fed" when evaluating processed feeds that are marketed in different ways, such as canned versus dry dog foods. *As-fed* refers to the foodstuff in the form in which it is fed to the animal, with the water included. The relationship between as-fed and dry matter will vary depending on the foodstuff, and how it has been processed.

Ether extract   The dissolving of the fat-based components in the feedstuff with ether. Ether extracts not only nutritional fat, but also other lipid-based materials in the plant. The value of this analysis varies with the type of plant tested.

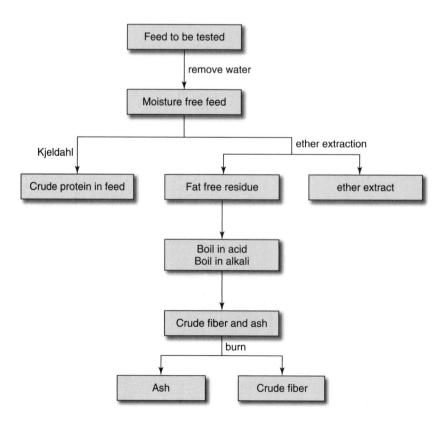

FIGURE 5–3 Proximate analysis of feedstuffs

$$\text{Nutrient digestibility (\%)} = \frac{\text{nutrient consumed} - \text{nutrient excreted}}{\text{nutrient consumed}} \times 100$$

FIGURE 5–4 Nutrient digestiblity calculation

Nitrogen-free extract (NFE) Carbohydrates that are not fiber. NFE is calculated by subtracting all of the other components determined by proximate analysis from the total amount of feed at the onset of the test. The result is the estimated NFE.

Fiber in feed can be analyzed with the **Van Soest analysis.** This form of analysis is best used with forages. This chemical analysis determines the acid detergent fiber and the neutral detergent fiber in the forage:

Acid detergent fiber The part of the feed that is most undigestible. This includes cellulose and lignin, as well as undigestible protein.

Neutral detergent fiber The material found in the cell walls of plants that does not break down during digestion in the simple stomach, but can be somewhat utilized by ruminant animals.

In addition to chemical analysis, feeds can be analyzed through **feeding trials.** A basic feeding trial involves feeding the feedstuff to the animal, and evaluating parameters such as growth, lactation, production, and so on. A digestion trial is conducted to determine how much of the nutrients in the feed the animal is using. After completion of a chemical analysis, the feed is given to the animal. Fecal material is collected and tested for the nutrient component of interest. The difference between the nutrients fed, and those excreted can be divided by the total intake and multiplied by 100 to give the digestibility of a nutrient (see Figure 5-4).

NUTRITION-RELATED TERMS

Anabolism (ahn-nahb-o-lihzm) The process of building complex compounds from simple compounds.

Body condition scoring A systemic process of evaluating the amount of fat on an animal's body. The scoring system varies with species. Body condition scoring is important to maintain the animal at the optimal condition. Higher numbers on the scale indicate more fat deposits; lower numbers indicate less fat deposits. The scoring ranges from **emaciated** (ē-mās-ē-ā-tehd), or extremely thin, to **obese** (ō-bēs), or extremely fat.

Calorie (cahl-oh-rē) A measurement of the amount of energy in a feed.

Catabolism (kah-tahb-o-lihzm) The process of breaking down complex compounds into their constituent parts.

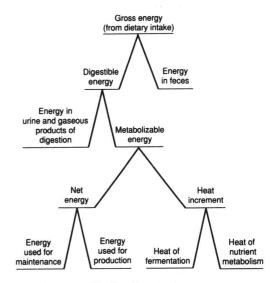

FIGURE 5–5 Energy utilization in animals

Deficiency (dē-fihsh-ehn-sē) An inadequate amount of a nutrient provided in the diet.

Dry matter intake The calculation of how much dry matter an animal can consume based on the animal's body weight. In most species, dry matter intake is based on 3–4 percent of the animal's weight.

Energy The portion of the feed available to power the body and its functions, and to generate body heat. Energy consumed in feed is used by the body in a variety of ways (see Figure 5–5). Dietary energy is classified in the following ways, based on how it is used:

Gross energy The total amount of energy in the feed or ration.

Digestible energy The gross energy minus the amount of energy that is not utilized in the body and is excreted through the feces.

Metabolizable energy (meh-tahb-oh-līz-ah-bl) The gross energy minus all other energy lost through urinary, fecal, or gaseous excretion.

Net energy Metabolizable energy minus the amount of energy expended to produce body heat. Net energy can further be defined by its use by the body, maintenance, production, lactation, and so on. An animal with a **positive energy balance** is consuming more energy than it is using; an animal with a **negative energy balance** requires more energy than it is consuming.

Feedstuffs Food.

Metabolism (meh-tahb-oh-lihzm) The process of converting food into the energy needed to maintain the body and perform bodily functions.

Palatability (pahl-ah-tah-bihl-ih-tē) The tastiness of the food. If the food is palatable, the animal will eat it. The most nutritious ration is of no value if the animal will not consume it.

Ration (rah-shuhn) The combination of feedstuffs that are provided to an animal, also may be called **diet.** A **balanced ration** is one that has the appropriate amounts of all of the nutrients that the animal needs.

Total digestible nutrients The total amount of nutrients in a feed that an animal can use.

Toxicity (tock-sihs-ih-tē) The presence of a chemical that leads to damage and destruction of cells. Toxicity can occur through overfeeding of some nutrients, from ingestion of plants that contain unhealthy components, or from consumption of contaminated feed.

TYPES OF FEED

Feedstuff are divided into two major categories: **concentrates** (kahn-sehn-trāts) and **forages** (for-ehj-ehs).

Concentrates

Concentrates are feed products that are relatively low in fiber and high in energy. Concentrates either provide energy in the form of carbohydrates, fats, or protein to the diet. Simple-stomached animals, such as swine and poultry, need high-carbohydrate diets because they cannot fully utilize the nutrients in high-forage diets. Feeds are often classified into the following groups based on the nutrients that they primarily provide in the diet:

Energy feeds These feeds have less than 20 percent protein, and provide energy through carbohydrates and fats. Most grains are energy feeds. Corn, oats (see Figure 5-6), and barley (see Figure 5-7) are common energy feeds. Energy feeds that are used in a particular region may vary depending on what grains are commonly raised in that area. Corn and

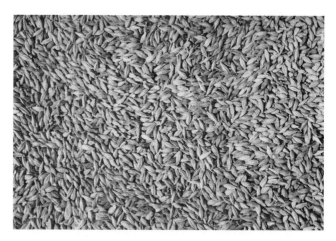

FIGURE 5–7 Barley, a cereal grain used in animal diets (Courtesy of ARS)

oats are commonly used throughout the country, but other grains may be used in regions where they are commonly raised.

Protein supplements Feeds that are high in protein, usually more than 20 percent. Protein supplements can either be from animal sources (meat, milk products, and so on), or plant sources (soybean oil meal, cottonseed oil meal, and so on). Soybean oil meal is one of the most often used protein supplements in animal feeds.

Forages

Forages are high-fiber feeds, and are also known as **roughage. Silage** (sī-lahj) is a type of forage that is commonly used for feed. Silage is a fermented form of a plant. Forages also include pasture and hay (see Figure 5-8), and are divided into **legumes** (lehg-yooms) and grasses:

Grasses A class of forages that is relatively low in protein. Common grasses used for animal feed are

FIGURE 5–6 Oats, a commonly used concentrate in horse diets

FIGURE 5–8 A round bale is a popular way to store hay, a common forage material

FIGURE 5–9 Orchard grass and the bloom of orchard grass. (Courtesy of Oregon State University, Forage Information System)

FIGURE 5–10 Alfalfa is a very popular forage, and is very high in protein.

timothy, orchard grass (see Figure 5-9), and Bermuda grass.

Legumes   A type of plant that can convert nitrogen into protein, using **nodules** (nohd-ūls) on their roots. Because they can do this, they are high in protein, up to 18 percent in some plants. Common legumes used for animal feed are alfalfa (see Figure 5-10), clover, and soybeans.

FACTORS AFFECTING NUTRITIONAL NEEDS

All animals do not have the same nutritional needs. Nutritional needs vary based on the type of activity the animal is involved in, and what the current demands are on the animal. The following terms are used to describe different levels of nutritional needs in animals:

Maintenance   The amount of nutrients needed to maintain basic bodily functions. All other categories of nutritional need are based on maintenance. The animal will use nutrients first to meet its maintenance requirements, and then additional nutrients will be used for other needs.

Growth   The amount of nutrients needed to maintain basic bodily functions, plus provide the nutrients needed for growth. The need for nutrients for growth is greatest in younger animals.

Lactation   The production of milk demands tremendous amounts of certain nutrients. A lactating female has one of the highest nutritional needs of any animal. In addition to an increased need for energy, lactating animals need more calcium and phosphorus than nonlactating animals.

Reproduction   Females need to be in optimal health to successfully become pregnant. As the fetus grows, there may be increases in the need for specific nutrients. In most species, the nutritional need does not increase until the last third of the pregnancy, when the majority of the fetal growth occurs. Cats are an exception, as they have increased nutritional needs from the beginning of the pregnancy. Females in early pregnancy that are lactating need to be fed to meet the lactational need, not fed to the reproductive need.

Work This category primarily affects dogs and horses. Animals that use energy and are involved in physical work will have an increased nutritional need when compared to animals that are not working. Energy for work will come after the animal has met its needs for maintenance; therefore, environmental characteristics such as temperature, can affect how much energy is needed for the working animal. For example, sled dogs in the Arctic will have a higher need for energy than dogs doing similar work in a temperate climate.

CHAPTER SUMMARY

Food and nutrition of animals is a primary area of focus in animal science. Understanding how the digestive tract of each species functions is important to maintain their well-being, as well as to maximize their production. All animals use the same basic nutrients; however, variations in the digestive tract lead to variations in feedstuffs used for individual animals.

STUDY QUESTIONS

1. What are the five classes of nutrients?

2. In what part of the digestive tract is pepsin added to the digesta?

3. In what part of the digestive tract do horses ferment roughages?

4. _____ are the forage source with the highest protein.

5. _____ is the classification for energy need on which all other classifications are based.

6. List three feedstuffs that qualify as concentrates.

7. What nutrient is the most energy dense?

8. What part of the ruminant stomach is the site of bacterial fermentation?
   a. Rumen
   b. Reticulum
   c. Omasum
   d. Abomasums

9. What is a balanced ration?

10. Call a local feed provider and learn the cost of some common feedstuffs. What class of feeds is most expensive per pound, forages or concentrates?

11. Select a species of interest and research what amino acids are essential and nonessential for that species.

Chapter 6
Animal Behavior

Chapter Objectives

► Understand primary classifications of animal behavior

► Explain the importance of animal behavior in animal management

There are a variety of labels or perspectives on the role of animals in society. Most people are somewhere along the continuum between the belief in **animal rights,** that humans should not use animals for food, entertainment, or any other purpose; and a **utilitarian** belief, that humans should be able to use animals for anything they wish, regardless of any pain or suffering on the part of the animal (see Figure 6-1). Most people believe in the importance of **animal welfare,** and

believe that it is acceptable to use animals for food and entertainment, and other benefits, but that as stewards of the animals, humans have an obligation to meet the animals' physical and psychological needs, and ensure minimum pain and suffering throughout the animals' lives.

The study of animal behavior plays an important role in understanding how human activity and management impacts animals. This area is of growing interest around the world, and concerns about animal treatment have significantly affected policies and laws regarding how people treat and handle their animals, both in production agriculture and ownership of companion animals.

The following are several approaches to the study of animal behavior:

Behavioral ecology (bē-hāv-yər-ahl ē-kahl-əh-jē) The study of how the behavior of a species interacts with its environment. This environment is not only the physical environment, but also the social and biological environment.

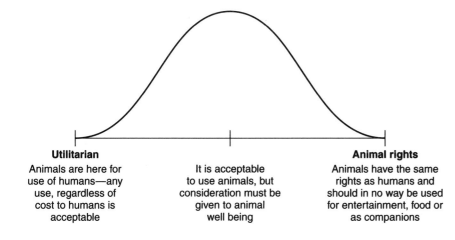

Utilitarian		Animal rights
Animals are here for use of humans—any use, regardless of cost to humans is acceptable	It is acceptable to use animals, but consideration must be given to animal well being	Animals have the same rights as humans and should in no way be used for entertainment, food or as companions

FIGURE 6–1 Continuum of attitudes toward human use of animals.

Comparative psychology (cahm-pār-ah-tihv sī-kahl-əh-jē) The study of the components that control behavior.

Ethology (ē-thohl-ah-jē) The study of an animal's behavior within its environment. This can be their natural environment, or the environment in which they currently live, such as a farm or zoo. The focus in ethology is often on learned or adaptive behaviors, and is less focused on the interactions between the animal and the environment than behavioral ecology.

Sociobiology The study of how behavior is related to biology, especially how behaviors may have developed to ensure species survival.

NORMAL BEHAVIOR

Many animal behaviors are rooted in the evolution of the animal, and are **innate** (ihn-nāt) behaviors that occur regardless of human intervention. It is important to realize that many behaviors that humans find frustrating in animals are normal behaviors for the animal. Changes in normal behavior are often one of the first signals that something is not right with the animal. The following are normal behaviors in animals:

Aggression Any behavior that poses potential danger or harm to another. Aggression can be either toward members of the same species (**conspecific**), or members of different species. Aggression can also be either **defensive,** to protect oneself, or **offensive,** to attack. It is important to understand the different types of aggression that animals may display. The following are types of aggression:

Fear aggression A type of defensive aggression. Fear aggression occurs when an animal feels it needs to be aggressive to protect itself. Many dog bites are based in fear-aggressive behavior. Animals that are fear-aggressive usually show body language that has a combination of submissive and aggressive characteristics, such as cowering in the corner and growling.

Maternal aggression Aggressive behavior resulting from a mother's instinct to protect her young. Maternal aggression usually decreases as the offspring becomes more independent.

Territorial aggression Aggressive behavior resulting from an animal's protection of its territory. Territorial aggression is seen most often in predatory-type animals, such as dogs. Territorial aggression often includes vocalization and offensive movements.

Possessive aggression Aggressive behaviors resulting from protecting a limited resource from others. Food protection is the most common display of possessive aggression. However, animals may also protect shelter or other resources that are valued.

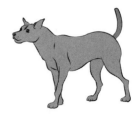

FIGURE 6–2 Dominance and submissive postures in dogs. Note that the dominant dog has a raised head and tail, whereas the submissive dog is dropping its head, lowering its neck, and has lowered its tail.

Critical period Developmental times in the lives of animals when they are most likely to retain experiences, either positive or negative, that will affect their behavior.

Dominance Relative ranking in the social order. Animals that exert dominance are higher-ranking than other animals. Some behaviors and postures are associated with dominance within the social structure (see Figure 6-2). For example, the dog in Figure 6-2 with the raised tail and raised head is in a dominant position.

Eliminative behavior Behaviors associated with elimination of bodily waste. Examples include burial of waste by cats, and urination and defecation to mark territory. Within species, different genders may demonstrate different eliminative behaviors.

Ethogram (ē-thō-gram) A record that indicates all of the behaviors that an animal exhibits in its environment.

Fear A natural response in any animal to anything that is considered a threat. Animals respond to fear by either **fight or flight,** standing and fighting off whatever is causing the fear, or fleeing from the fearful situation (see Figure 6-3). The choice to flee or fight varies with species and by situation. Even species that have a preference for flight will fight when threatened if they cannot escape.

Flight zone The distance to which an unknown individual can approach an animal before it flees (see Figure 6-4). Flight zones are largest in wild animals, and smaller in domestic animals.

Flocking/herding instinct The instinct in some types of animals to stay together in groups.

Hierarchy (hī-er-ahr-kē) Also known as "pecking order," hierarchy describes the structure of the social group relative to the degree of dominance. The order may be linear, or nonlinear, where more than one

FIGURE 6–3 A dog in a fearful posture. Note how closely the dog is pressed to the floor, and the lowered and extended head.

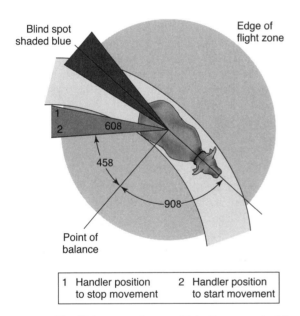

FIGURE 6–4 The flight zone of a cow. Note the present of the blind spot, and the positions where the handler can influence the animal's movement. (Courtesy Temple Grandin, Ph.D., 1989)

animal may have the same relative rank (see Figure 6-5).

Ingestive Behaviors associated with the consumption of food and water.

Predatory (prehd-ah-tōr-ē) Behavior associated with stalking and killing prey. Predatory behavior differs from aggressive behavior in that it is generally silent.

Sexual behavior Behavior related to the successful attraction of a mate and mating. Sexual behaviors vary greatly among species. Sexual behaviors include those that indicate that a female is receptive to breeding (see Figure 6-6), and those of a male attempting to attract the attention of a female.

Socialization The process of acclimating an animal to its environment, and to the wide range of stimulus it may experience later in life. Socialization is especially critical with companion animals that live in close contact with people. Socialization is most effective when it occurs early in an animal's life.

Social behavior The behaviors that are exhibited among animals as they interact with one another. In herd animals, social behaviors help solidify the structure of the herd. An example of an animal social behavior is the act of grooming one another.

Stimulus (stihm-yoo-luhs) Anything that causes a response in an animal. Novel stimuli are anything new or unusual with which the animal is not familiar.

Stress Any stimulus that result in a change in the body's normal state. Stress is not always a result of negative stimulus; positive stimulus can also result in increased stress for an animal.

Submission Behavior of an individual that is lower in the hierarchy, or behavior that indicates a lack of relative power. Young offspring often exhibit submissive behaviors toward their dams and toward older animals in the social group.

Temperament An individual animal's behavioral response to its environment. Temperament is a combination of genetics and previous experiences an animal has had.

ABNORMAL BEHAVIOR

Animals exhibit a wide range of abnormal behaviors. Determining the source of a behavior is vital to determining how to minimize or eliminate a behavior. Vices are abnormal behaviors that create management or health problems. Often, vices and abnormal behaviors develop as ways for animals to cope with stress. This stress often results from an inability to express normal behaviors. Abnormal behaviors that are repetitive and have no apparent purpose are known as **stereotypic** behaviors. The following are stereotypic behaviors:

Anorexia (ahn-ō-rehck-sē-ah) A refusal to eat.

Cannibalism The consumption of an animal of the same species that can occur in a variety of species. Young animals are most often the victims of cannibalism, although chickens will peck and cannibalize pen mates.

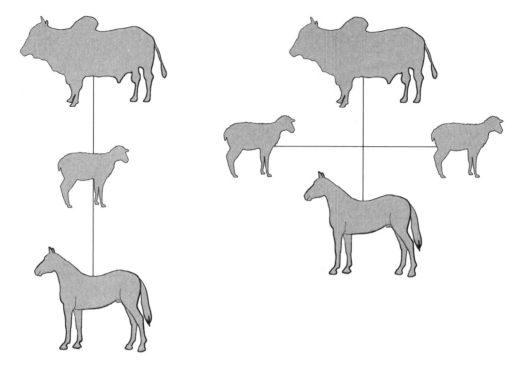

FIGURE 6–5 Hierarchal arrangements in animals can be either linear or non-linear.

FIGURE 6–6 Stallion and mare mating

FIGURE 6–7 Horse cribbing.

Cribbing A compulsive behavior in horses in which they grasp a solid object with their upper teeth and inhale air (see Figure 6-7). Cribbing is a very undesirable behavior that can lead to serious health problems, such as colic.

Coprophagy (kōp-rō-fah-jē) The eating of fecal material, either of the same species or of different species. In some young animals, coprophagy is normal as it inoculates the gastrointestinal tract with necessary microbes. However, coprophagy in adult animals is abnormal, and may be indicative of a lack in their diet.

Phobia (fō-bē-ah) Extreme fear of something.

Pica (pī-kah) Ingestion of inappropriate materials. Please note that pica is the consumption of materials that are not normal feedstuffs. For a dog, removing tasty scraps from the trash is a very normal behavior, not an example of pica.

Stereotypy (stār-ē-ah-the-pē) A repetitive, nonproductive behavior with no apparent purpose or intent. The following are examples of stereotypic behaviors:

Stall weaving In horses, this is a behavior in which a horse stands in a stall and shifts its weight back and forth from one front leg to the other. In extreme cases, the horse takes sideways steps with the front legs. Also known as **weaving.**

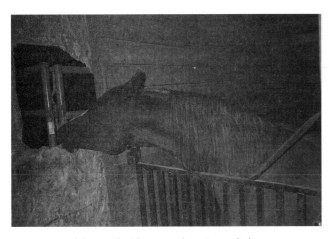

FIGURE 6–8 A horse that is a consistent wood chewer

Tongue rolling   Extension and rolling of the tongue for no apparent purpose.

Wood chewing   Often in horses, the consumption of wood fences or walls (see Figure 6-8). This differs from cribbing in that they don't inhale air.

Wool chewing   In sheep, chewing and eating the wool until they have bald spots.

Wool sucking   A behavior in cats during which they imitate the nursing and kneading behaviors of kittens after weaning.

BEHAVIORAL MODIFICATION

Behavioral modification refers to efforts to mold animal behavior to better suit human purposes. Behavioral modification is also known as training. Behavioral modification has a growing place in animal management as people strive to minimize stress in the lives of animals. Behavioral modification is very important for companion animals and horses because of the close interaction between humans and the animals. The following are terms related to behavior modification:

Aversive (ah-vers-ehv)   A negative stimulus that the animal will wish to avoid in the future.

Clicker training   A method of behavioral modification using **operant conditioning** that links the click of a plastic cricket with the promise of a reward. The click can then be used to identify desirable behaviors, and stimulate repetition of those behaviors as the animal seeks the reward. This method of training was first developed for use with marine mammals, and has become very popular for use with dogs, horses, and other species.

Conditioning   The learning of a behavior in response to a stimulus. The following are different methods of conditioning:

Classical   The linking of a stimulus to an action, resulting in a physical response. A classic example is how Pavlov linked a ringing bell to feeding dogs, and eventually, the ringing bell stimulated salivation without the presence of food.

Operant   The linking of a stimulus with a response. Rewards are often used to create the link between a stimulus and a desired response. The reward needs to be something that the animal values. Food is a common reward, but some animals are more driven by play or other rewards than by food.

Counterconditioning   Teaching a behavior that is incompatible with the undesirable behavior. For example, if a dog runs barking at the door every time someone knocks, teach the dog to lie on its bed every time there is a knock at the door. Running to the door and lying on the bed are incompatible behaviors. If the dog associates lying on its bed with receiving a desired award, the dog will select the desired behavior.

Extinction   Eliminating a behavior by not reinforcing it. It is important to note that any attention for a behavior provides reinforcement. When combined with attention for desirable behaviors, extinction can effectively eliminate some undesirable behaviors. For example, if a dog jumps on people, turning away and ignoring the dog provides no reinforcement. If the dog is ignored for jumping, but praised and petted when all feet are on the floor, the dog will quickly learn that not jumping gains attention, and jumping gains no attention. Extinction is only effective if the behavior is ignored; punishing or negatively reinforcing a behavior is not extinction. Punishment is a form of attention.

Habituation   Also called **desensitization,** habituation involves an animal becoming sufficiently accustomed to a stimulus that no longer elicits a response. For example, if a train runs past a pasture of cattle, the cattle may flee the first few times, but eventually they learn that the train is not a threat, and ignore the train when it passes.

Imitative   A behavior learned from copying another animal. Animals learn many behaviors as offspring, by imitating the behaviors of their mothers or littermates.

Imprinting   A type of learning that occurs in an animal's first few hours of life. Imprinting is how animals identify their mothers and other individuals in their social group. In training some species, handlers use imprinting to introduce the animal to common management practices.

Learned   A behavior that is not naturally occurring, but has been gained through behavior modification.

Punishment   Modification of a behavior by creating an aversive situation when the behavior occurs. Punishment is the least effective method of behavior modification because it must occur instantly, and every time the behavior occurs to be effective. In addition, punishment may lead to fear of the individual giving the punishment, or may reinforce the animal's fear of a situation. If a horse is afraid to approach a puddle, punishing the horse will only reinforce the belief that the puddle is a danger.

Sensitization   Increasing the response to a stimulus. When an animal is sensitized to a stimulus, decreasing intensity of the stimulus will result in a desired behavior. For example, sensitization can occur in training an animal with leg cues when being ridden. With training, an animal can become so sensitized to the stimulus that it is barely discernible.

Reinforcement   A use of positive or negative actions to increase or decrease a behavior. **Positive reinforcement** is giving an animal something desirable to encourage repetition of a behavior. **Negative reinforcement** takes away something bad when a desired behavior occurs. For example, a rider puts a spur in a horse's side to encourage it to move, and then removes the spur from the side when the animal moves away.

    Intermittent reinforcement   Once a desired behavior is learned, it no longer needs to be reinforced every time it occurs. A schedule of reinforcing the behavior occasionally is known as intermittent reinforcement. Intermittent reinforcement can be at a fixed or variable interval. Fixed interval is when the behavior is reinforced every X times it occurs, and variable interval is when there is no set number of times that behavior must occur to be reinforced. Intermittent reinforcement is a very strong reinforcer of behaviors.

Successive approximation   Also known as shaping behavior, successive approximation modifies behavior by rewarding animals as they make attempts toward the desired behavior. As successive approximation proceeds, the animal must perform a behavior more and more similar to the desired behavior. For example, when teaching a cow to enter a milking parlor, the cow is rewarded each time it gets closer to the parlor. Eventually, the cow has to enter the parlor to be rewarded. Successive approximation is an extremely effective way to overcome fear of places or objects. It is important with successive approximation to reward every bit of progress toward the desired goal to keep the animal trying to achieve the desired behavior.

COMMUNICATION

Animals communicate in many ways. A good animal handler needs to be aware of this communication, and pay attention to the messages. It is very important to look at all of the messages that an animal sends, as drawing conclusions from one or two pieces of information may lead to an inaccurate interpretation of the signals. Research indicates that children who grow up with animals are more socially successful upon entering school; one theory is that those children have learned to read body language and other nonverbal communication through their interactions with animals.

Auditory Communication

Auditory (aw-dih-tōr-ē)   Communication is vocal. Each species makes unique sounds. The variation of those sounds is a method of communication. The following list includes some of the sounds each species makes.

    Bovine
      Moo (moo)   Basic vocalization used to call offspring, or to communicate with herd mates.
    Canine
      Bark   Sharp vocalization that can announce the presence of a newcomer or indicate excitement. Wolves have a much higher bark threshold than domestic dogs, and are much less likely to bark.
      Bay   A deep-throated howl-like sound in some breeds of hounds, primarily those that hunt by scent. Baying occurs when the prey is located.
      Growl   A low and deep vocalization in the chest. Usually indicates either defensive or offensive aggression.
      Grunt   A sound of satisfaction and contentment.
      Howl   Believed to be a way for dogs to find their pack mates. Wolves often howl if they become separated from their pack.
      Whine   Relatively high-pitched vocalization that usually signals discomfort or distress, such as hunger, loneliness, or attention seeking.
    Feline
      Growl   A low, sound deep in the chest indicating aggression.
      Hiss   A defensive vocalization that occurs when the cat is feeling threatened. In extreme cases, a hiss may be accompanied by spitting (see Figure 6-9).

FIGURE 6–9 A hissing kitten.

Mating call A loud, carrying vocalization that is used to attract the attention of males. Cats are solitary animals, so they must announce their reproductive status to males.

Meow The basic feline sound that seems to be used primarily to communicate with people, not with other cats.

Purr A sound made deeply in the chest. Purring is generally associated with pleasure and contentment, but cats will also purr when sick or injured.

Olfactory Communication

Olfactory (ohl-fahck-tōr-ē) Communication is very important to animals. Many animals identify each other by smell. Animals also mark their territories by defecating, urinating, or scratching to release scents into an area. Scents left by an animal give other animals information on the species, gender, and reproductive status.

Body Language

Body language is also an important form of communication for animals. To anticipate how animals will react to a situation, it is important to understand the messages they send. An understanding of animal body language allows safer and more productive interactions with animals. Each part of the animal's body can be used to communicate. To understand what the animal is communicating, all of the information the animal is providing must be gathered. The following are examples of body language:

Bunting A behavior in cats in which they rub their heads on objects to leave a scent mark. Cats exhibit this behavior on stationary objects, and on people.

Ear position Ear position provides important information. An animal with ears up and forward is paying attention to what the ears are pointing toward (see Figure 6-10). If the ears are flat to the head, the animal is agitated.

Piloerection (pī-lō-ē-rehck-shuhn) Commonly called "raised hackles," the raising of the hair over the neck and shoulders often accompanies growling, barking, or hissing.

Striking Most often seen in horses, striking is kicking out with the front leg. This behavior indicates irritation or upset, and may also accompany sexual behavior.

FIGURE 6–10 The ear position of horses communicating different messages.

CHAPTER SUMMARY

Animal behavior is a growing field of research in the animal sciences. Researchers are studying how management affects behavior, and making recommendations for management to maximize the well-being of animals. In addition, understanding animal behavior plays a vital role in our interactions with animals. We often wish to modify the behavior of animals, either to a great degree, such as with training horses and dogs, or to a moderate degree in regard to routine handling of production animals. Behaviors can be successfully modified with any animal if the animal's natural responses are considered, and reinforcement is timely and consistent. When seeking to modify behaviors, it must be remembered that the reinforcement must occur within three seconds of the behavior being modified. It is especially important to have impeccable timing if using punishment or the wrong behaviors will be reinforced, and the animal may become fearful and more difficult to handle. In most cases, rewarding positive behaviors will result in a longer-lasting, more consistent change in the behavior of the animal than attempting to punish for undesirable behaviors.

STUDY QUESTIONS

1. What are the major approaches to the study of animal behavior?

2. What is a stereotypy? Give an example.

3. How does successive approximation modify behavior?

4. Why is it difficult to modify behavior successfully using punishment?

5. What can you learn from an animal's body language?

6. What message is a cat sending when its ears are flattened to its head?

7. What is anorexia?

8. On the Internet, find pictures of a species of your choice exhibiting different body language. Print the pictures and describe the message the animal is sending.

9. Make an ethogram of your day.

Chapter 7
Animal Disease and Parasites

Chapter Objectives

▶ Define disease

▶ Understand the basics of disease transfer

▶ Understand how to prevent disease

▶ Identify diseases in animal species

▶ Learn the diseases that pose a risk to human health

▶ Learn the parasites that affect animals and how to control them

Disease (dih-zēz) is a term that broadly describes any difference from the normal health of a living thing. Diseases can include metabolic diseases, such as **ketosis** (kē-tō-sihs), injuries such as broken bones and cuts, or diseases caused by other organisms, such as bacteria, viruses, and parasites. Some diseases are also genetic, and are inherited from the parents.

Infectious diseases are caused by pathogens that multiply in the body and cause damage to the body. When the number of pathogens in a body reaches a high concentration, disease can be passed on to others. Diseases can be passed from one animal to another in a variety of ways. Some diseases are passed on by **vectors,** another living being that carries the disease from one animal to another. Mosquitoes and ticks are two insects that serve as vectors. Disease can also be passed directly from animal to animal through blood, saliva, or other

secretions. Some diseases are transmitted by coughing and sneezing, when the pathogen is carried on droplets of moisture through the air. Other pathogens can live in the soil or on surfaces for extended periods of time, and be passed to animals that come in contact with those surfaces.

Although most diseases affect a specific species, some diseases can be passed from animals of one species to animals of another. Diseases that can be passed from animals to humans are **zoonotic** (zoo-oh-naht-ik) diseases, and are of special concern to animal health and public health officials.

The study of diseases is **etiology** (ē-tē-ohl-ō-jē). **Etiologists** study the causes and transmission of, and the cures and preventions for diseases. **Symptoms** of disease are the characteristics that affected animals show and that make a person believe the animal has a disease. Some diseases have similar symptoms, which makes them difficult to identify or diagnose.

DISEASE PREVENTION AND TREATMENT

Disease prevention and treatment are important parts of animal ownership and management. The following are some basic steps to prevent and treat disease:

Antibiotic (ahn-tih-bī-oh-tihck)   A chemical given to an animal to battle disease organisms. Antibiotics are very specific to the disease organism in question. If an animal is being raised for consumption, it is important to be aware of the **antibiotic withdrawal time,** the time that must pass between the treatment and sale of the animal. Observing antibiotic withdrawal time ensures that antibiotics used in animals are not introduced into the food supply.

Antibody (ahn-tih-boh-dē)   An antibody is produced by an animal's immune system to fight off disease-causing organisms. Specific antibodies are produced to battle specific disease-causing organisms.

Antigen (ahn-tih-jehn) Something that the body believes is foreign, and that stimulates a response from the immune system. Bacteria and viruses are types of antigens.

Biosecurity A plan or process to prevent disease organisms from entering a facility or a country.

Core vaccine A vaccine recommended for all animals.

Recommended vaccine A vaccine recommended for animals fitting a set of guidelines that indicate they are at risk for the disease.

Resistance This occurs when a disease-causing organism or parasite develops an ability to withstand chemicals designed to protect animals. Resistance can develop either to antibodies, or to the chemicals used to fight parasites.

Sanitation The practice of keeping a facility clean and free of contaminants. Sanitation is a crucial step to reducing disease in a herd or flock of animals.

Titer (tī-tər) The measure of antibodies in the blood of an animal. Measuring the titer is one way to determine whether an animal has sufficient defense against an organism.

Vaccination (vahck-sihn-ā-shuhn) The introduction of a form of a disease-causing organism to an animal to stimulate the body to produce antibodies to fight the organism. Vaccines can be **modified live,** which means that the disease-causing organism has been changed to make it less dangerous, or **killed,** in which case the disease-causing organism is dead and cannot reproduce. Some vaccines come in both forms. Also available are some **recombinant vaccines,** which are genetically modified to stimulate a response from the immune system without causing disease. A veterinarian should be involved in developing a vaccination program for animals.

COMMON DISEASES

Infectious diseases (ihn-fehck-shuhs dih-zēz) are those that are caused by pathogens (pahth-ō-jehnz), which are microorganisms. Pathogens are often **bacteria** and **viruses,** although other organisms can cause disease, and are most often species-specific. For example, multiple species, including humans, can get influenza; however, the disease-causing pathogen for each species is species-specific. The pathogens causing most common diseases are **endemic** (ehn-dehm-ihck) in the environment, meaning they are always around.

Different things can result in an animal contracting a disease. Animals that are under stress are more likely to contract a disease than those that are not. Stress results in a decrease of an animal's ability to resist disease. Poor nutrition, or a lack of needed nutrients, can cause stress that weakens an animal's **immune** (ihm-yoon) system. The immune system is a body's defense system against infection. Diseases can be either **chronic** (krohn-ihk), which means they are of moderate intensity, last for a long time, and may come and go; or can be **acute** (ah-kūt), which means they are very intense and relatively short in duration. Some diseases can begin as an acute disease, and then become a chronic disease.

Noninfectious diseases are those that are caused by something other than a pathogen, and are not passed from one animal to another. Noninfectious diseases include nutritional diseases, genetic diseases, and allergies. In some diseases, multiple factors, such as genetics and nutrition, can contribute to the expression of the disease.

INFECTIOUS DISEASES

Following is a list of common infectious diseases, all caused by some type of pathogen:

Abscesses Abscesses are an infection that the body contains in a pocket. These are often caused by bacteria. In cats, abscesses are usually associated with injuries resulting from fighting, and are often seen in male cats that are frequently outdoors. Draining abscesses and treating with antibiotics is effective. Species affected: all species

Anthrax (ahn-thracks) Anthrax is caused by bacteria that live in the soil, and can live in a dormant state for many years. Animals are infected by consuming the bacteria when grazing, or by infection from another animal through a vector, such as a biting insect. Symptoms include fever, **ataxia** (ā-tacks-ēah)—the loss of muscular control—difficulty breathing, and eventually collapse (see Figure 7-1). An effective vaccine can be used to control anthrax in areas of concern. Species affected: bovine, equine

Arthritis An inflammation of the joints that can be caused by either bacteria or a virus. Arthritis results in swollen joints and lameness. Affected birds may lose weight and may have diarrhea. Some bacterial forms of arthritis are treatable with antibiotics, but the viral form is not. Species affected: avian

FIGURE 7–1 A cow that died of anthrax (Courtesy of USDA)

Aspergillosis (ahs-per-jeh-lō-sihs) A type of pneumonia caused by a fungus or a mold. Aspergillosis can affect birds of any age, and has a high mortality rate in young birds. Symptoms include loss of appetite and loss of weight. Eliminating exposure to the fungus or mold is the best preventative; there is no effective treatment. Species affected: avian

Atrophic rhinitis (ā-trō-fik rī-nī-this) Two different bacteria, *Pasteurella* or *Bordetella,* can infect the nasal passages and cause atrophic rhinitis. Symptoms include sneezing, tearing, and nasal discharge. As the bones atrophy, the snout becomes twisted. Species affected: porcine

Avian erysipelas (air-ih-sĭp-ə-lehs) An infection caused by bacteria that most commonly affects turkeys. Birds on the flock often begin dying before other symptoms of diarrhea and anorexia are noticed. Treatment with antibiotics may be effective, and birds can be vaccinated if erysipelas is present in the region. Species affected: avian

Avian influenza Caused by a virus, avian influenza can affect birds, causing coughing, sneezing, and weight loss. Avian influenza is being closely monitored due to concerns that a mutation of the virus may create a human health hazard. Species affected: avian

Avian leukosis (loo-kō-sihs) Marek's disease and lymphoid leukosis are two forms of avian leukosis, and are caused by different viruses. Marek's disease affects the nerves and results in paralysis and death; lymphoid leukosis affects the bones. There is no treatment for either form of the disease, but a vaccine does exist for Marek's disease. Species affected: avian

Blackleg A disease caused by **anaerobic** bacteria. Anaerobic bacteria are those that do not require oxygen to survive. The bacteria that causes blackleg live in the soil and enter an animal's body either by mouth, or by entering through a wound. The bacteria infect the wound and cause lameness, depression, and fever (see Figure 7-2). Blackleg can be treated with antibiotics if it is diagnosed early. An effective vaccine does exist for blackleg. Species affected: bovine

Bluecomb This term refers to two different diseases. The turkey form is caused by a virus and results in diarrhea and dehydration. Although birds of all ages can be affected, bluecomb is most fatal for young birds. Bluecomb in chickens is caused by a different organism than that in turkeys, but is probably still a virus. Birds become anorexic, decrease production, and lose weight. The comb may also become bluish in color. Antibiotics can be used to treat bluecomb in turkeys and chickens. Sanitation is important in preventing both forms of bluecomb. Species affected: avian

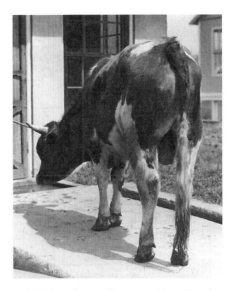

FIGURE 7–2 Blackleg disease in a cow. Note the characteristic swelling of the right hind leg. (Courtesy of USDA)

Bluetongue Caused by a virus and spread by gnats, blue tongue results in loss of appetite, and swelling around the head and mouth. Vaccination is an effective preventative. Treatment consists of treating the symptoms. Affected species: caprine, ovine

Bordetellosis (bōr-deh-tah-lō-sihs) Also known as kennel cough, bordetellosis is a bacterial upper respiratory infection that affects dogs and cats. A vaccine is effective, and is highly recommended for animals that are exposed to a large number of other animals, such as through showing or boarding. Species affected: canine, feline

Bovine respiratory syncytial virus (sihn-sihsh-uhl) A bacterial disease that affects the respiratory system, primarily in young calves. Stress increases the likelihood of this disease. An effective vaccine does exist. Species affected: bovine

Bovine spongiform encephalopathy (BSE) (spuhn-ghi-form ĕn-sĕf-ah-lōp-ah-thē) Also known as "mad cow disease," BSE affects the central nervous system of cattle and other ruminants. A federal surveillance plan is in place in the United States to detect any signs of BSE in the U.S. cattle population. BSE is believed to be transmitted through consumption of feed that contains prions. Prions are only present in the central nervous tissue in an infected animal, so the risk of transmission to humans is extremely low. Symptoms of BSE include nervousness, weight loss, and aggression. There is no vaccine or effective treatment for BSE. Species affected: bovine

Bovine viral diarrhea (BVD) A disease that affects cattle of all ages. The symptoms include fever, coughing, slow weight gain, and diarrhea. An affected pregnant cow may abort the fetus, or deliver a mummified fetus, depending on the stage of pregnancy during which the cow became infected. An effective vaccine does exist for BVD. Species affected: bovine

Brucellosis (broo-sehl-ōh-sĭhs) A serious disease in cattle that is caused by a microorganism that also causes undulant fever in humans. Brucellosis can result in sterility in cows or bulls, late-term abortion, and delivery of weak calves. Brucellosis can be spread through direct contact with an infected animal, an infected fetus, or with feed or water that has been infected. There is no cure for brucellosis, but effective vaccination and screening programs have greatly reduced the incidence. Because of the disease's tremendous economic impact, state and federal programs assist in the elimination of brucellosis in cattle. Species affected: bovine, canine, porcine

Calf enteritis (scours) *Scours* is a broad term for calf diarrhea. The causative organism will vary depending on the part of the country. *E. coli* is one of the common bacteria that causes scours. Scours most often affects young calves. In acute cases, calves may have cold extremities (ears, nose, legs) and diarrhea, and may die quickly after symptoms are noticed. In chronic cases, calves have diarrhea and lose weight until death occurs. Sanitation of all items that the calves had contact with will reduce the incidence of scours. It is also important that affected calves receive antibody-rich first milk, **colostrom** (kō-lah-struhm), from the cow. Vaccination of cows prior to calving increases the antibodies in the colostrum, and provides additional protection for calves. Species affected: bovine

Canine distemper (dis-tehm-pər) Caused by a virus and transmitted through the virus in secretions of the infected animal, distemper can cause anorexia, coughing, and sneezing. In severe cases, the central nervous system is affected. Distemper can result in a high death rate in unvaccinated puppies. Vaccination is an effective preventative and is a core vaccine in dogs. Species affected: canine

Canine parvovirus (pahr-vō-vī-rhus) A highly infectious viral infection that results in diarrhea. Parvovirus can also damage the heart of affected dogs. Parvovirus is most often seen in puppies, and has a high mortality rate in unvaccinated puppies. Parvovirus is a core vaccine in dogs. Species affected: canine

Chlamydia (klah-mĭd-ē-ah) An infection with the *Chlamydia* bacteria can cause conjunctivitis (infection of the eyelid), or upper respiratory infection in cats. Antibiotics are an effective treatment. Species affected: feline

Cholera (kahl-er-ah) Historically a devastating disease with a large number of hogs dying annually, cholera has been eradicated in the United States, although it still occurs in other countries. Hog cholera is caused by a virus, and characterized by loss of appetite and fever. Species affected: porcine

Clostridial diarrhea (klah-strihd-ē-ahl) A bacterial infection of the intestine that causes diarrhea in pigs less than a week old. The mortality rate may be as high as 25 percent. Symptoms are watery diarrhea and lethargy. Sanitation of facilities is the best prevention, and vaccination of sows may be helpful. Treatment of infected piglets is usually unsuccessful. Species affected: porcine

Encephalomyelitis (ehn-sĕf-ah-lō-mī-ah-lī-tihs) Also known as sleeping sickness, encephalomyelitis refers to several strains of a virus that affects the brain. All strains of encephalomyelitis are carried by mosquitoes, and are characterized by fever, depression, circling, and pressing of the head against a solid surface. Vaccination and elimination of mosquito breeding areas are effective in prevention. Species affected: equine

Enterotoxemia (ehn-tār-ō-tock-sēm-ē-ah) Also known as overeating disease, enterotoxemia is caused by bacteria and affects goats and sheep. Vaccination of lambs and ewes, and ensuring adequate roughage in the diets of lambs will reduce the incidence. Once the symptoms have begun, treatment is not available. In cattle, enterotoxemia can result from overeating concentrated diets. Vaccination can also prevent symptoms in cattle. Species affected: bovine, caprine, and ovine

Equine infectious anemia (EIA) Also known as swamp fever, EIA is caused by a virus that biting insects such as flies and or mosquitoes carry. There is no cure, or effective treatment for EIA. Horses can be tested for EIA with the **Coggins test,** which detects the existence of antibodies to equine infectious anemia. Many states require documentation of a negative test for EIA for horses to be transported across state lines, or within the state. Species affected: equine

Equine protozoal myelitis (EPM) (prō-tō-zō-ahl mī-eh-lī-tihs) EPM is caused by protozoa that infect the central nervous system. Opossums are an intermediate host, and infection commonly occurs when feedstuffs, such as hay, are contaminated with infected fecal material. Horses with EPM show symptoms typical of neurological challenge, and may be uncoordinated. Treatment with antibiotics has shown some success, and a vaccine is available, although the effectiveness of the vaccine has not been proven as of publication of this book. Species affected: equine

FIGURE 7–3 A pig infected with swine erysipelas (Courtesy of USDA)

Erysipelas (air-ih-sihp-ə-lehs) Caused by bacteria, erysipelas can affect pigs at any age between weaning and maturity. Animals with erysipelas have red lesions on the skin, and develop arthritis that leads to lameness (see Figure 7-3). Vaccination is an effective prevention for erysipelas. Species affected: porcine

Exudative epidermitis (ehcks-yoo-dā-t ə hv ehp-ih-dər-mī-tihs) Occurs when bacteria on the skin infect areas where the skin has been broken by bites, scratches, or other injuries. The bacteria multiply, and the animal develops red spots that ooze fluid. Eventually, the animals may become dehydrated from loss of the fluid. Antibiotics can be effective for treatment under the supervision of a veterinarian. Reducing stress and maintaining a clean environment will prevent the disease. Species affected: porcine

Feline calicivirus (kah-lēk-ē-vī-ruhs) A viral upper respiratory infection in cats that can be transmitted through direct contact between animals. Symptoms include sneezing, fever, and anorexia. Vaccination is effective in prevention, and treatment consists of treating the symptoms while the virus runs its course. Feline viral rhinotracheitis is a similar disease. Calicivirus and rhinotracheitis are core vaccines for cats. Species affected: feline

Feline immunodeficiency virus (FIV) (ihm-myoo-nō-dē-fish-ə hn-cē) FIV is caused by a lentivirus that attacks the immune system of the cat. Most often, FIV is transmitted through bites from infected cats. Preventing exposure to infected animals is important. FIV is fatal, and there is no cure. Although similar to human immunodeficiency virus (HIV) in humans, FIV is species-specific, and is not transmissible to humans. Species affected: feline

Feline infectious peritonitis (FIP) (pehr-ih-tō-nī-tihs) FIP is caused by a coronavirus, and symptoms include anorexia, fever, vomiting, and diarrhea. Diagnosis can be confirmed through laboratory testing. There is no effective treatment. Vaccination is

available, but is not widely used. Species affected: feline

Feline leukemia virus (FeLV) (loo-kēm-ē-ah) FeLV is caused by a virus and can be transmitted from cat to cat through secretions such as saliva. Biting is not necessary to transfer the virus. Fever, depression, and anorexia are common symptoms. Vaccinations do exist, and high-risk cats (those that spend significant time outside, live with cats with FeLV, and so on) should be vaccinated. There is no effective cure for the disease, and animals may carry the virus and show no symptoms. Species affected: feline

Feline viral rhinotracheitis (FVR) (rīnō-trāk-ē-ī-this) FVR is a viral upper respiratory infection in cats caused by a herpesvirus that can be transmitted through direct contact between animals. Symptoms include sneezing, fever, and anorexia. Vaccination is effective in prevention, and treatment consists of treating the symptoms while the virus runs its course. Similar to feline calicivirus. Species affected: feline

Foot and mouth disease (FMD) Foot and mouth disease is caused by a virus. Although humans can transmit the virus to animals on their clothes, and can carry the virus in their nasal passages for a few hours, FMD is not transmissible to humans, and does not pose a human health risk. Reproductive efficiency, milk production, and growth all decrease when an animal is infected with FMD. A vaccination does exist; however, vaccinated animals can spread the disease. Because of the significant economic impact of an outbreak of FMD, some countries choose to destroy infected animals to minimize the opportunity for an outbreak. Species affected: bovine, caprine, ovine, deer

Foot rot A general term used to describe an infection of the foot that can be caused by a virus, bacteria, or fungus. These pathogens enter the foot through a break in the skin, and cause an infection that results in lameness. The animal may get a fever and lose weight, especially if the lameness makes it difficult to access feed. Foot rot can be prevented by maintaining a clean environment for the animal, and can be treated with antibiotics. Species affected: bovine, caprine, ovine

Infectious bovine rhinotracheitis (IBR) A viral disease with several forms, depending on what part of the system the virus affects. **Respiratory IBR** affects the respiratory system and is the most common form. **Genital IBR** affects the genitalia of either males or females. **Conjunctival IBR** (kohn-juhnck-ti-vahl) affects the eye, and **encephalitic IBR** (ehn-sehf-ah-lih-tik) affects the brain and nervous system. IBR may cause abortions in pregnant animals. No treatment exists for IBR; however, vaccination programs and quarantine of new animals can minimize the incidence of IBR. Species affected: bovine

Influenza A respiratory disease caused by a viral (in the case of swine flu, also a bacterial) infection. Vaccination can prevent the disease, and most animals recover, although the disease results in decreased growth. Antibiotics are not effective in treatment. Species affected: canine, equine, porcine

Johne's disease (yō-nēz) Also known as **paratuberculosis** (pahr-ah-tuh-berk-yoo-lō-sihs), Johne's disease is a chronic disease that is characterized by diarrhea in adult animals. Diarrhea causes weight loss and decreased production in animals, and may lead to death (see Figure 7-4). There is no effective treatment for Johne's disease, and care should be taken to avoid bringing affected animals into a herd. Species affected: bovine, ovine

Mastitis (mahs-tī-tihs) An inflammation of the mammary gland. Mastitis can either be caused by a pathogen, or can be noninfectious in nature. Antibiotic treatment is effective, but recurring mastitis can cause permanent damage to the mammary gland. A gland with mastitis may be swollen, hot, and sore. If the female is nursing young, the presence of mastitis will reduce the ability of the offspring to obtain milk. Milk from dairy cows that have mastitis, or are being treated with antibiotics for mastitis, cannot be sold for human consumption (see Figure 7-5). Sanitation is vital to the prevention of mastitis. Species affected: All mammals.

Mastitis-metritis-agalactia (MMA) (ā-gahl-ahck-tē-ah) A complex disease of unknown cause, MMA is characterized by a sow having no milk (agalactia)

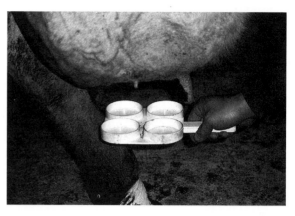

FIGURE 7–5 The California Mastitis Test (CMT) is used to determine if a cow has a mastitis infection. (Courtesy Ron Fabrizius, DVM, Diplomate ACT.

when piglets are born. The mammary gland shows signs of mastitis. Reducing stress on the sow appears to minimize incidence of MMA. Species affected: porcine

Metritis (meh-trī-tihs) An infection of the uterus that can be caused by a variety of pathogens. Symptoms include depression, loss of appetite, and failure to conceive. Animals that retain the placenta after parturition are more likely to have metritis. (A retained placenta is when a placenta does not completely disengage from the uterus, and all or part of the placenta remains in the reproductive tract.) Treatment with antibiotics can be effective for metritis; however, it is important to identify the pathogen that causes the infection for effective treatment. Species affected: all mammals

Mycoplasmosis (mī-kō-plahz-mō-sihs) A group of respiratory diseases in poultry that are caused by a variety of *Mycoplasma* bacteria. Different bacteria cause different strains of the disease. The disease can spread rapidly through a flock and result in significant economic loss. Treatment and prevention varies with the strain of bacteria present. Species affected: avian

Mycotoxicosis (mī-kō-tocks-ih-kō-sihs) A broad term used to describe disease caused by fungus or molds in the feed that are ingested by birds, and then produce toxins in the body. Symptoms range from anorexia to death, depending on the level and type of toxin present. The best preventative is the proper storage of feed to eliminate exposure to molds and fungus. Species affected: avian

Navel Ill A bacterial infection of the navel that occurs shortly after birth. Causes joint swelling and pain, weight loss, and potential death. Antibiotics are an effective treatment, and good sanitation practices prevent infection. Species affected: all mammals

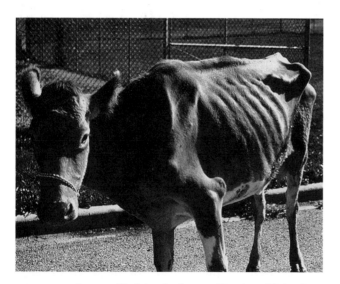

FIGURE 7–4 A cow with Johne's disease (Courtesy Michael T. Collins, DVM, PhD, University of Wisconsin School of Veterinary Medicine)

Necrotic enteritis (neh-krah-tik)   Caused by bacteria that inflames the intestinal tract and causes diarrhea and anorexia. Treating with antibiotics may be helpful, as well as separating sick and healthy pigs, and maintaining sanitary conditions. Species affected: porcine

Newcastle disease   A viral disease of major concern in the poultry industry. Several viruses can cause different forms of Newcastle disease. Affected birds have difficulty breathing, appear nervous, and may show neurological signs such as tremors and paralysis. Birds in production will show a decrease in production. There is no treatment for Newcastle disease, but sanitation and vaccination can reduce the incidence. Species affected: avian

Pasteurellosis (pahs-cher-ehl-ō-sihs)   An infection in rabbits caused by the bacteria *Pasteurella multocida.* The bacterial infection can become localized in different areas, resulting in diseases with different names. **Snuffles** is a common form of pasteurellosis, which is concentrated in the upper respiratory tract (see Figure 7-6). *Pasteurella* can cause **pneumonia,** an infection of the lungs; **pyometra** (pī-ō-mē-trah), an infection in the uterus; **orchitis** (ōr-kī-tihs) an inflammation of the testicles; **otitis** (ō-tī-tihs) **media,** an infection of the middle ear; **conjunctivitis,** an inflammation of the eye; **abscesses** in the skin; and **septicemia** (sehp-tih-sē-mē-ah), a blood infection. Species affected: rabbits

Pinkeye   An infection of the cornea of the eye caused by bacteria. The bacteria can be carried by insects, by dust, or directly transferred from one animal to another. If pinkeye is left untreated, it can permanently damage the eye and cause blindness. Animals with white faces and pink skin around the eyes are more likely to be affected. The symptoms include pinkness and cloudiness of the eye, swelling, and tearing. Some vaccinations are available, and a veterinarian should be consulted to determine the best vaccine. Cattle can be treated with antibiotics in the eye, and by protecting the affected eye from light. Species affected: bovine, caprine, ovine

Pneumonia (nū-mō-nē-ah)   An inflammation of the lungs, pneumonia has many potential causes: bacterial, viral, and parasitic. Pneumonia often accompanies another illness that has weakened the immune system and stressed the body. Antibiotics can be effective for treatment. Severe cases can cause permanent damage to the lungs, even if the animal survives. Species affected: all

Potomac horse fever (PHF)   Characterized by diarrhea, fever, and depression, PHF is caused by the organism *Ehrlichia risticii.* The organism cannot live outside a cell, so cannot be transmitted through the environment. The precise mechanism of spread is unclear, but it appears that some type of vector, probably insects, is involved. Secondary diseases, such as laminitis, are also of great concern in horses with PHF. Treatment with antibiotics is successful, and horses in areas with a high occurrence of PHF can be vaccinated. Species affected: equine

Pseudorabies (soo-dō-rā-bēz)   A viral disease that can affect pigs of any age. The symptoms vary somewhat with the age group. In general, affected animals have fevers, vomiting, and may cease feeding. Antibiotics are not effective for treating pseudorabies. Various hog-raising regions have eradication plans in place for pseudorabies. Pigs should be purchased from certified pseudorabies-free herds. Species affected: porcine

Rhinopneumonitis (rī-nō-noo-mah-nī-tihs)   An upper respiratory infection caused by any of several herpesviruses. Vaccination is effective for preventing rhinopneumonitis. Rhinopneumonitis can cause late-term abortions in pregnant mares. A killed virus vaccine has been developed to protect pregnant mares from the virus. Species affected: equine

Scrapie (skrā-pē)   In the same family of diseases as bovine spongiform encephalopathy, scrapie attacks the central nervous system. Scrapie has an incubation period of a year, so is most often seen in older animals. Symptoms include weight loss, abnormal walking gait, and loss of coordination (see Figure 7-7). There is no cure or treatment, and the USDA has a voluntary program in place to identify scrapie-free flocks in an effort to reduce the incidence of the disease. Affected species: caprine, ovine

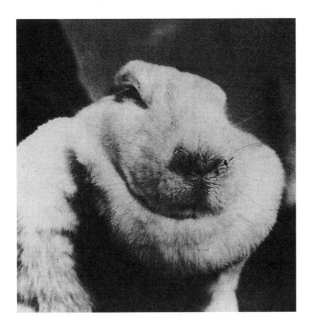

FIGURE 7–6 A rabbit with Pasteurellosis in the upper respiratory tract (snuffles). (Courtesy of USDA)

FIGURE 7–7 A sheep with scrapie. Loss of wool and drooping ears are consistent with an advanced stage of the disease. (Courtesy of USDA)

Shipping fever Shipping fever is a broad term for a complex disease that is caused by stress, a viral infection, and a bacterial infection. The disease complex results in respiratory symptoms, such as difficulty breathing, coughing, and nasal discharge. Animals are depressed, do not eat, and have diarrhea, which results in significant weight loss. Affected animals are susceptible to secondary infections, which result when the body is too weakened to fight them off. Vaccination and reduction of stress are the best preventatives. Treatment with antibiotics can be effective if begun early in the progression of the disease. Species affected: bovine

Sore mouth A viral infection that results in pus-filled blisters in affected areas. The lips, mouth, and udder are most affected. Blisters create discomfort when eating, so many animals lose weight or don't allow young to nurse if infected (see Figure 7-8). Vaccination can prevent sore mouth. Affected species: caprine, ovine

Strangles (distemper) In horses, strangles is caused by the bacteria *Streptococcus equi.* Symptoms include swelling of the lymph nodes under the jaw, nasal discharge, and lethargy. A vaccine is available, and a veterinarian should help determine if the vaccine is appropriate for a given situation. Antibiotics can help in treating the disease. Species affected: equine

Swine dysentery (dihs-əhn-tər-ē) Also known as bloody scours, swine dysentery is caused by bacteria. Symptoms include anorexia and slow weight gain. The bacteria are shed in fecal material, and can be carried on unclean clothing and tools from farm to farm, or barn to barn. Antibiotic treatment can be effective, but disease prevention is a priority. Species affected: porcine

Tetanus (teht-nuhs) Also known as lockjaw because of the characteristic rigidity of the jaw in affected animals, this bacteria enters the body through open wounds. Affected animals move stiffly and eventually have muscular spasms and die. There is no

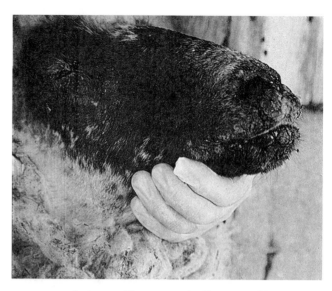

FIGURE 7–8 A sheep with soremouth. (Courtesy of Ron Fabrizius, DVM, Diplomate ACT)

effective treatment for tetanus, but vaccination is very effective. Affected species: bovine, caprine, equine, ovine, human

Vibriosis (vihb-rē-ō-sihs) A bacterial infection that causes abortions. Vaccination is an effective preventative, and any aborted fetuses should be destroyed to reduce contamination. Affected species: bovine, caprine, ovine

Warts Skin growths caused by a virus. They most often affect young animals, and can be easily spread from animal to animal. Animals with warts should be separated from others to prevent spreading the virus. Species affected: bovine, equine

NONINFECTIOUS DISEASES

A variety of diseases are not caused by specific pathogens. The following is a list of common noninfectious diseases:

Allergies Allergies result when the immune system overreacts to the presence of a substance, usually a protein, that the body identifies as foreign. Although any species can have allergies, they are of greatest concern in dogs, cats, and horses.

Anhydrosis (ahn-hī-drō-sihs) This is not a disease caused by a pathogen, but is a condition in which the animal's sweat glands do not function at all, or function abnormally. Horses with anhydrosis have difficulty controlling their body temperature, especially when working. Species affected: equine

Canine hip dysplasia (CHD) (dihs-plā-zē-ah) A syndrome in dogs in which the head of the femur does not fit tightly into the acetabulum of the pelvis. This

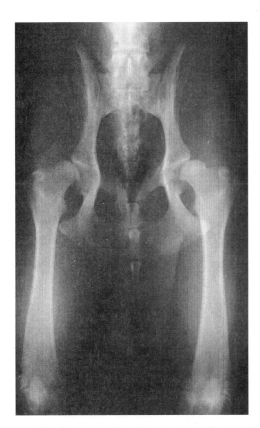

FIGURE 7–9 A radiograph of the hip structure of a dog. Note how the femoral head fits into the acetabulum. In a dog with hip dysplasia, the femoral head does not fit snugly in the acetabulum. Radiographs such as this can be used to diagnose hip dysplasia. (Courtesy of Lodi Veterinary Hospital, Lodi, WI)

FIGURE 7–10 Classic laminitic stance in a horse. Note how the horse is keeping the weight off the front feet.

results in arthritis in the joint, and pain and lameness in the dog. CHD is a polygenic (multiple-gene) syndrome that is exacerbated by accelerated growth and/or excessive exercise in young dogs. Radiographic evaluation of the hips should be conducted on animals selected for breeding (see Figure 7-9). Species affected: canine

Laminitis (lahm-ih-nī-tihs)   A complex disease with many potential causes. Laminitis is the inflammation of the sensitive laminae in the hoof of the animal. Overfeeding, overexercising, stress, systemic infections, and some medications can cause laminitis (see Figure 7-10). Laminitis can result in permanent damage to the hoof and may limit an animal's ability to move comfortably. If laminitis is suspected, a veterinarian should be contacted immediately, especially in the case of horses, where a reduction of athletic ability could result in an inability to serve their purpose. **Founder** is a term often used interchangeably with laminitis. *Founder* is more appropriately the term used to describe the

chronic condition following an acute bout of laminitis, from which there is permanent damage to the feet, and often malformation of the hooves. Species affected: bovine, equine most often. Could occur in any hooved animal.

Navicular disease (nah-vihck-yoo-lahr)   A complex syndrome that occurs when the navicular bone in horses degenerates, causing pain and lameness. Horses susceptible to navicular disease are typically relatively large horses with disproportionately small hooves. Navicular disease can be managed with medications and proper shoeing, but cannot be cured. Species affected: equine

Periodic ophthalmia (of-thahl-mē-ah)   Also known as moonblindness, periodic ophthalmia is an inflammation of the eye that results in swelling and cloudiness. Deficiency of vitamin B is linked to moonblindness. The horse may lose some or all of its vision. Periodic ophthalmia is a chronic condition, and the symptoms may recur periodically. Species affected: equine

Porcine stress syndrome (PSS)   A genetic disease in hogs. Affected animals are easily excited and die due to problems with the circulatory system when stressed. The most effective management practice is to avoid introducing the PSS genes into the herd. Meat from animals with PSS is of less desirable quality than from normal pigs. Species affected: porcine

Recurrent airway obstruction (RAO)   This disease is commonly known as **heaves.** Symptoms include difficulty breathing, either when exercised, or in severe cases, when resting. Horses with RAO often show a **heave line,** which is an exaggeration of the muscles below the ribs caused by the stress of inhaling and exhaling. Species affected: equine

Torticollis (tōr-tih-kō-luhs)   A condition in rabbits where the muscles near the cervical vertebrae contract,

FIGURE 7–11 A rabbit with torticollis (wry neck). (Courtesy of USDA)

causing a tilting of the head (see Figure 7-11). Also known as wry neck, torticollis can be a result of pasteurellosis, parasitic infection, or injury. Species affected: rabbits

NUTRITIONAL DISEASES

A wide range of diseases have a nutritional or metabolic component. The most common causes are excess or deficiency of particular nutrients that disrupt the balance of an animal's bodily functions. Nutritional disease can be widely prevented with a well-balanced diet of high-quality ingredients. The following are examples of nutritional diseases:

Anemia (ah-nēm-ē-ah)   A lack of iron in the diet causes anemia, which results in poor growth, reduced immune function, and potentially, death. Anemia can be treated by supplementation of iron. Anemia can also result from blood loss due to parasite infestation. Species affected: all

Azoturia (ahz-ō-tur-ē-ah)   Because azoturia is characterized by stiffness and soreness when returning to work after time off, it is also known as Monday-morning sickness because it was prevalent in working horses on Monday mornings after having Sunday off. Reducing carbohydrates in the diet when not working reduces the incidence of azoturia. Species affected: equine

Bloat (blōt)   A disease in ruminants that results from fermentation in the rumen, causing excessive gas. Most often, bloat occurs when too much green grass is consumed. Large-breed and deep-chested dogs are also susceptible to bloat from the excessive production of gas in the stomach. Species affected: bovine, canine

Botulism (botch-yoo-lizm)   A form of food poisoning, also known as limberneck, caused by consuming feedstuffs that contain the botulism bacteria. The bacteria result in paralysis and death. An antitoxin can reverse the affects of the toxin, but is generally not cost-effective for an entire flock. Species affected: avian

Chronic obstructive pulmonary disease (COPD)   Also known as **heaves** (hēvz), COPD is similar to asthma and is a respiratory disease that results from consuming dusty or moldy hay. Animals with COPD have difficulty breathing, and must use their abdominal muscles to exhale, resulting in a muscular heave line that can be observed along the barrel. Minimizing exposure to dust is the best treatment for COPD. Species affected: equine

Colic (kohl-ihck)   A broad term used to describe a variety of conditions affecting the digestive tract of horses. The primary symptom of colic is abdominal pain. Colic can be caused by sudden changes in the diet, parasites, diets of poor-quality feed or insufficient forage, or dehydration. Colic symptoms can result from excessive gas (**gas colic**), blockage of the digestive tract (**impaction colic**), or twisting of the intestines. Veterinary intervention is usually required to resolve colic. Feeding a diet high in forages, and maintaining a regular deworming program can reduce the incidence of colic. Species affected: equine

Displaced abomasum (DA)   Occurs primarily in dairy cattle, when the abomasum shifts in the abdomen, usually after calving. DA is most common in cattle on a high-concentrate diet. Feeding roughages and gradually increasing concentrates to meet nutritional needs is the best prevention. Surgical intervention is sometimes needed to correct the displacement (see Figure 7-12). Species affected: bovine

Eclampsia (ē-klahm-sē-ah)   Also known as **milk fever,** eclampsia is a loss of muscle control and weakness that is a result of calcium depletion in the body. Eclampsia can be treated with calcium supplementation. Species affected: all mammals

Fatty liver syndrome   A disease affecting birds on a high-concentrate diet. Birds get diarrhea and anemia, and may die with no apparent symptoms. Species affected: avian

Fescue toxicosis   Feeding animals fescue, a type of grass, creates health problems in some groups. Cattle can get **fescue foot** (see Figure 7-13), lameness in the hind feet that may result in hoof loss, from grazing on tall fescue. The exact mechanism of spread is not understood. Pregnant and lactating mares should not eat fescue because an **endophyte** (ehn-dō-fīt), which is a parasite in the grass, can cause thickening

FIGURE 7–12 Surgery for a displaced abomasum in a cow.

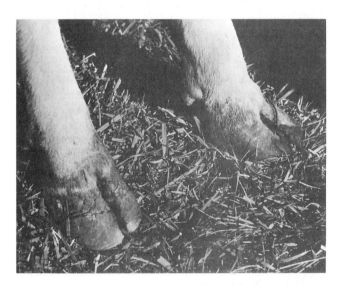

FIGURE 7–13 The hoof of a cow affected by fescue foot. Notice the characteristic splitting of the hoof. (Courtesy of USDA)

FIGURE 7–14 Treating a cow with intravenous calcium-magnesium gluconate for grass tetany. (Courtesy of USDA)

of the placenta and agalactia. Species affected: bovine and equine

Grass tetany (teht-ahn-ē)   Occurs when cattle are fed a diet that is deficient in magnesium, one of the trace minerals. Animals affected are usually fed on grass that is grown on magnesium-deficient soil, and can be treated by supplementing magnesium (see Figure 7-14). Species affected: bovine

Hardware disease   Occurs when cattle ingest metal objects when eating. The metal becomes lodged in the reticulum. Sharp objects may puncture the reticulum and cause injury to internal organs. Removal

of metal objects that may be ingested is the best prevention. Cattle magnets are available, which are consumed and rest in the reticulum and attract metal objects. The magnet helps keep the metal from puncturing the reticulum. Species affected: bovine

Hypoglycemia (hī-pō-glī-sēm-ē-ah)   A condition that results when blood sugar levels are extremely low. This is usually due to insufficient sugar in the diet. Treatment is supplementation with simple sugars, and increasing the sugar in the diet. Species affected: canine, porcine

Ketosis (kē-tō-sihs)   A metabolic condition that occurs when carbohydrates in the diet are insufficient to provide for the animal's energy needs. Ketosis is characterized by weight loss, lethargy, and the smell of acetone on the animal's breath. Species affected: bovine, specifically high-producing dairy cattle

Night blindness   A lack of vitamin A results in the loss of vision, weakness, and potentially convulsions. Adequate vitamin A in the diet prevents the disease. Species affected: caprine, ovine

Parakeratosis (pāhr-ah-kāhr-ah-tō-sĭs)   A condition that results from zinc deficiency. Zinc deficiency can be either from lack of zinc in the diet, or an imbalance of calcium and zinc. Zinc supplementation controls the condition. Species affected: porcine

Plant toxicity   A wide variety of plants produce toxic compounds that can cause illness or death. These toxic compounds can be found in any part of the plant, and different toxins are present at different times of year. Toxic plants must be removed from any areas where animals graze on pasture, or where hay is harvested. Prevalence of plants is very regional, and different plants affect different animals. Species affected: all species

Rickets (rihck-əhts)   Symptoms are slow growth and crooked development of bones. Rickets can be caused by an imbalance or deficiency of calcium and phosphorus. Lack of vitamin D, which is essential for use of calcium, may also result in rickets. Species: all

Urinary calculi (yoo-rihn-ār-ē kahl-kyoo-lī)   Small stones from the crystallization of minerals in the urine that form in the urinary tract. The stones may block the urethra, preventing the flow of urine from the bladder. Males are most often affected. In cats, the formation of stones in the urine is known as feline lower urinary tract disease. Adjustments to minerals in the diet can prevent the formation of calculi. Species affected: bovine, caprine, ovine, feline

White muscle disease   A disease caused by deficiency of the mineral selenium. White muscle disease is characterized by weakness and lack of muscle control. Animals can be treated with selenium supplementation, but it is important to remember that selenium toxicity can also cause illness. Species affected: bovine, caprine, ovine. Feline and canine can be affected, but this seldom occurs.

ZOONOTIC DISEASES

Zoonotic diseases are diseases that can be transmitted from animals to humans. They can be transmitted either directly between animal and human, through a vector, such as a mosquito or tick, or through the environment. Diseases that are zoonotic become of concern not only to animal scientists and those in the animal industry, but also to those concerned with public health. The following are examples of zoonotic diseases:

Borreliosis (Lyme disease) (bōr-ehl-ē-ō-sihs)   A bacterial disease that is transmitted to humans and dogs through ticks. Lameness and joint swelling are common symptoms in dogs. Treatment with antibiotics is effective. A vaccine is available, and is used in dogs that engage in high-risk activities, such as hunting in areas where borreliosis is endemic. Species affected: canine, human, equine

Brucellosis   Brucellosis is caused by different bacteria in dogs than in cattle, but has similar symptoms of abortion and reduced fertility, with very few other definable symptoms. Brucellosis is transmitted through sexual activity, and testing of breeding animals is recommended. Transmission of the bacteria to humans who handle infected fetal material is possible, although highly unlikely. Treatments are available, but outcomes are variable. Species affected: canine, human

Cat scratch disease   The bacteria that causes cat scratch disease is endemic in the cat population. Cats do not show symptoms from the bacteria; however, humans can become infected following a scratch or bite. Human symptoms include fever, pain, nausea, and enlarged lymph nodes. People with compromised immune systems should avoid contact with cats that may scratch or bite. Species affected: feline, human

Leptospirosis (lehp-tō-spī-rhō-sihs)   Caused by bacteria, leptospirosis can be transmitted between different species and to humans. The bacteria are transmitted through urine or through contaminated soil and water. There are several strains of leptospirosis, and vaccines are strain-specific. Work with a veterinarian to vaccinate for the strain most prevalent in certain geographical locations. Animals may be **asymptomatic** (ā-sīhmp-tō-mah-tihck), and show no symptoms, or may have a fever, stiffness, and loss of appetite. Species affected: bovine, canine, porcine, human

Listeriosis (lĭs-tear-ē-ō-sihs)   Caused by a germ, listeriosis can be spread through contaminated feed or water, and by contaminated bodily fluids. **Encephalitic** (ehn-sehf-ah-lih-tihck) **listeriosis** affects the brain, and **septicemic** (sehp-tih-sēm-ihk) **listeriosis** affects the whole body. There is no vaccine for listeriosis, and the disease often progresses to death. Sanitation is the best preventative. Species affected: most mammals and birds

Plague   Caused by the bacteria *Yersinia pestis,* plague in cats has two forms, bubonic and pneumonic. The bacteria are present in populations of rodents that cats hunt and consume. Treatment with antibiotics can be affective. Care must be taken, especially with pneumonic plague, to prevent spread to humans. Species affected: feline, human

Rabies (rā-bēz)   Rabies is caused by a virus that can affect any mammal. Wild animals, such as bats, skunks, and raccoons, provide a reservoir for the disease organism. Rabies can be transmitted through the saliva of an infected animal when it bites another animal. Rabies is a major human health concern, and veterinarians may recommend vaccinating horses in areas where rabies in the wild population is high. Affected animals show nervous and aggressive behavior and excessive salivation. There is no cure. Species affected: all mammals

Salmonellosis (sahl-muh-nehl-ō-sihs)   Infection with the *Salmonella* bacteria results from consumption of the bacteria through food, or exposure to the bacteria through infected materials. *Salmonella* infection results in vomiting and diarrhea. Antibiotics can be effective for treatment; sanitation is the best preventative. Salmonellosis can be transmitted from animals to humans, and is a primary zoonotic disease of concern. Species affected: all

Toxoplasmosis (tocks-sō-plahz-mō-sihs) A disease caused by a protozoal parasite in cats. Symptoms include fever, diarrhea, muscle weakness, and other neurological signs. Pregnant women can become infected with the parasite by cleaning a litter box of infected cats. Infection can result in birth defects, so pregnant women should not handle cat fecal material, or clean litter boxes. Species affected: feline

Tularemia (too-lah-rēm-ē-ah) Primarily affects cats that hunt outdoors and consume rodents that harbor the bacteria. It can also be contracted through ticks. Symptoms include fever, mouth ulcers, and eventually damage to the liver and bone marrow. Infected animals often die, and can transmit the disease to humans through bites or bodily fluids. Species affected: feline

West Nile encephalitis (WNE) (ehn-sehf-ah-lī-tihs) A relatively new disease in the United States, West Nile encephalitis is carried by mosquitoes that have bitten a bird that carries the WNE virus. The infected mosquito then bites an animal and infects the animal with the virus. Symptoms are similar to other diseases that affect the central nervous system: anorexia, circling, and loss of coordination. A vaccine does exist for WNE, and treatment can be effective. Species affected: canine, equine, human

PARASITES

Parasites are any of a variety of organisms that live by "stealing" nutrients from their hosts. Parasites can be categorized as internal or external, based on where they attack the host. They can either carry other diseases, or cause damage to the body directly. Animals with heavy loads of parasites can show symptoms of malnutrition due to the quantity of nutrients the parasites are taking. Animals that are infested with parasites grow more slowly and produce less than animals without parasites. Some parasites can cause permanent damage to affected organs.

Parasites can be treated with a variety of medications that kill or remove the parasites. **Anthelmintics** (an-thehl-mihn-tihcks) are medications that are used to kill internal parasites. Understanding the life cycle of different parasites, and breaking that life cycle, is also effective in controlling parasites.

Ectoparasites

Ectoparasites (ehck-tō-pahr-ah-sīts), or external parasites, live on the outside of the body of an animal. They are vectors for a variety of diseases, often cause skin irritation, and may cause anemia due to the

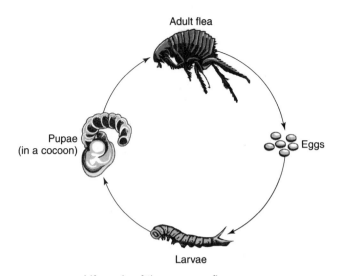

FIGURE 7–15 Life cycle of the common flea

amount of blood they consume. Ectoparasites can often be identified visually with the naked eye. The following are some ectoparasites:

Fleas An insect that sucks blood from the host. A heavy infestation of fleas can remove enough blood to result in anemia, and even death, for the host (see Figure 7-15). Fleas are large enough to be visually seen. Also, the "flea dirt" or excrement, may be noted, especially in the inguinal region. Fleas can also be intermediary hosts for other diseases and parasites. Some animals, especially dogs, have allergic reactions to flea saliva when they are bitten. Although fleas will bite humans, they do not stay on human hosts. There are a variety of treatments and prevention programs for fleas. Consult with a veterinarian, and ensure that products are approved for use in the species of concern. Species affected: dogs, cats, humans

Flies A wide variety of flies affect livestock. Danger from flies comes from two sources: transmission of the disease through biting, and irritation of the animal that increases stress and decreases growth and production. Flies can be treated with insecticides, and by eliminating breeding areas. Species affected: all species

Lice Blood-sucking insects that affect a variety of species (see Figure 7-16). Lice infestations can be prevented by isolating new animals and ensuring they are not infested before exposing the whole herd. Topical treatments are effective for eliminating lice. Species affected: all species, by species-specific varieties

Maggots The larvae of insects, usually flies, that can live in animal tissues. A maggot infestation results when they lay eggs on an animal, usually in an area that is wet or where the skin is damaged, and the

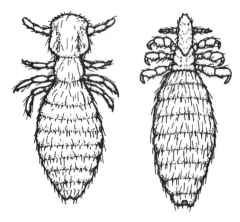

FIGURE 7–16 Two kinds of lice that attack sheep. Left, a biting louse; right, a sucking louse (Courtesy of University of Illinois at Urbana-Champaign)

FIGURE 7–17 A cow with scabies caused by mites. (Courtesy of USDA)

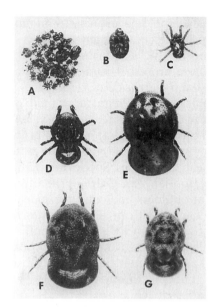

FIGURE 7–18 The bovine ear tick. A-ticks and debris collected from ear; b-engorged larva; c-young nymph; d-partially engorged nymph; e-fully engorged nymph; f-adult female; g-adult male. (Courtesy of Texas Agricultural Extension Service, Texas A&M Univeristy)

eggs hatch and feed on the tissue. Species affected: all species

Mites Small spider-like insects that suck blood from their hosts. A variety of mites affect different species of animals (see Figure 7-17). Mites cause **mange** (mānj), a skin disease characterized by itching, redness, and hair loss. The two major varieties of mange are **sarcoptic** (sahr-kohp-tihck) and **demodectic** (deh-mō-dehck-tihck), each caused by mites of the same name. Sarcoptic mange is characterized by severe itching caused by the mites burrowing under the skin. The severe itching can result in abrasion of the skin, leaving it susceptible to secondary infection. Demodectic mange is less severe than sarcoptic mange, and is characterized by more mild, localized, itching. Because mites bite their hosts, they can also spread some diseases. Species affected: all species

Mosquitoes One of the most dangerous external parasites are mosquitoes. Although the quantity of blood they remove is minimal, mosquitoes are vectors for a wide variety of diseases and other parasites that they can spread to most animal species. Elimination of breeding ground is the best way to minimize mosquito infestation. Species affected: all species

Ringworm An infectious skin disease caused by a fungus that can be easily spread to other animals and to humans. Ringworm is characterized by a round patch of hairless skin, and can occur anywhere on the body (see Figure 7-19). Topical disinfection of affected sites is an effective treatment. Affected animals should be kept separate from nonaffected animals to minimize spread. Sanitation is important to prevention of ringworm. Species affected: all species

Ticks Several species of ticks affect animals. Ticks are blood-sucking insects that also carry organisms that cause other diseases (see Figure 7-18). Species affected: all species

Endoparasites

Endoparasites (ehn-dō-pahr-ah-sīts) are parasites that live in the body of the host during their infective stage. Most endoparasites are commonly known as "worms," or "internal parasites." Although the infective stage is internal to the host, many endoparasites spend at least some part of their life cycle outside the body of the host. The following are some of the endoparasites that affect animals:

Ascarids (ahs-kah-rids) Ascarids (large roundworms) are major parasites of concern in animals. Different

FIGURE 7–19 A cow with ringworm.

species of ascarids affect different species of animals. Mature worms in the intestines release eggs that are shed in the feces, and then are ingested by an animal that is grazing or pecking in the grass. The eggs hatch in the digestive tract, then the larvae invade the walls of the intestinal tract and migrate through the body. Some larvae end up in the lungs, where they are coughed up and swallowed again to mature in the intestinal tract, shed eggs, and continue the cycle. Some larvae become encysted in the tissue of the host animal. In the case of dogs and cats, pregnancy can cause the reactivation of these encysted parasites, and they can be transmitted to the young through the placenta in dogs, or through the milk in cats (see Figure 7-20). Species affected: canine, equine, feline, porcine

Bots (bahts)   Parasites that live in the stomach of horses during the larval stage. Bot flies lay yellowish eggs on the horse (the location depends on species of fly), that then ingests the eggs that hatch and mature in the stomach. The mature larvae are passed out in the feces, where they complete maturation to the fly stage, and the cycle begins again (see Figure 7-21). Species affected: equine

Coccidiosis (kohck-sihd-ē-ō-sihs)   A protozoal parasite that causes irritation of the intestinal tract, leading to diarrhea that is often bloody. Animals do not eat well, lose weight, and may die. Coccidiosis can be treated with antibiotics, and many starter feeds for birds contain coccidiosis medication. Different species of the parasite affect different species of animals and birds, so it is important to use the medicated feed for the correct species. Species affected: all livestock and avian species

Gapeworms   A parasite that lives in the trachea of birds, and can interfere with breathing. Birds ingest parasite-infected snails or worms, so range-fed birds are at the most risk of infection (see Figure 7-22). Species affected: avian

Heartworms   Heartworms affect both dogs and cats, although dogs are the primary hosts. Heartworm is transferred when a mosquito bites an infected dog, picks up the larval worm, and then passes it to another dog through a bite. The worms lodge in the heart to mature, and cause exercise intolerance and eventually death (see Figure 7-23). Laboratory tests can detect the presence of heartworms, and there are effective treatments. Excellent preventative

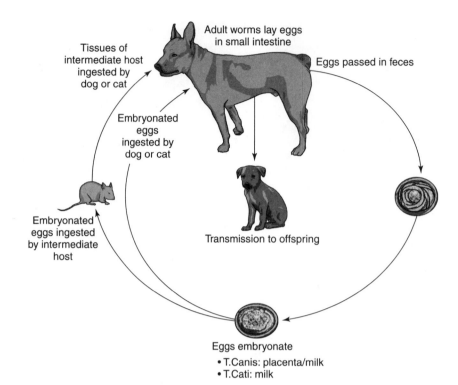

FIGURE 7–20 The life cycle of the ascarid

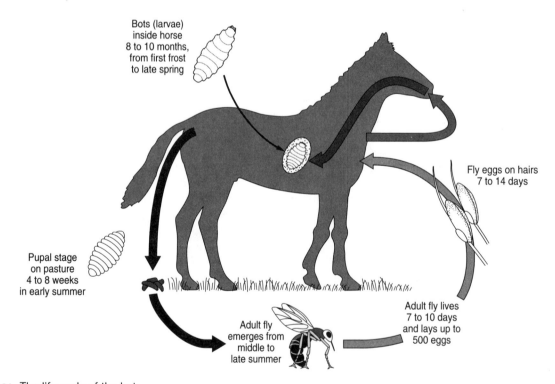

FIGURE 7–21 The life cycle of the bot

medications are available, and it is recommended that dogs in places where heartworm is endemic are maintained on a preventative. Species affected: canine, feline

Liver fluke A parasite that lives in the liver. The eggs are shed through the manure, and the larvae then

hatch. The larvae mature in an intermediate host, the snail. After maturing, the larvae leave the snail and are ingested again when the larvae attach to blades of grass. Species affected: caprine, ovine

Lungworm Thread lungworms and hair lungworms live in the lungs. Eggs are laid in the lungs, coughed

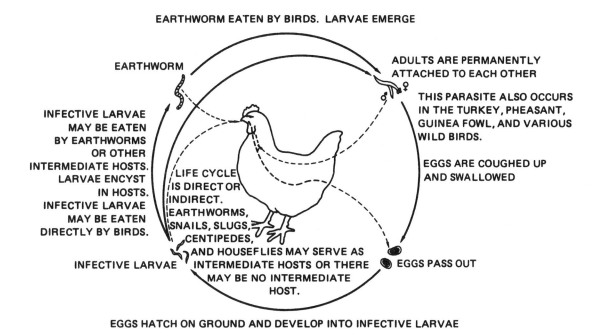

EARTHWORM EATEN BY BIRDS. LARVAE EMERGE

EARTHWORM

ADULTS ARE PERMANENTLY
ATTACHED TO EACH OTHER

INFECTIVE LARVAE
MAY BE EATEN
BY EARTHWORMS
OR OTHER
INTERMEDIATE HOSTS.
LARVAE ENCYST
IN HOSTS.
INFECTIVE LARVAE
MAY BE EATEN
DIRECTLY BY BIRDS.

THIS PARASITE ALSO OCCURS
IN THE TURKEY, PHEASANT,
GUINEA FOWL, AND VARIOUS
WILD BIRDS.

EGGS ARE COUGHED UP
AND SWALLOWED

LIFE CYCLE
IS DIRECT OR
INDIRECT.
EARTHWORMS,
SNAILS, SLUGS,
CENTIPEDES,
AND HOUSEFLIES MAY SERVE AS
INTERMEDIATE HOSTS OR THERE
MAY BE NO INTERMEDIATE
HOST.

INFECTIVE LARVAE

EGGS PASS OUT

EGGS HATCH ON GROUND AND DEVELOP INTO INFECTIVE LARVAE

FIGURE 7–22 The life cycle of the gapeworm (Courtesy of University of Illinois at Urbana-Champaign)

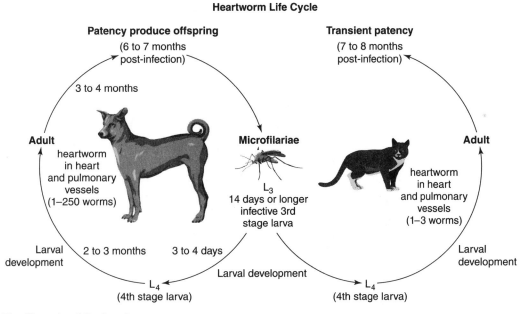

Heartworm Life Cycle

Patency produce offspring
(6 to 7 months
post-infection)

Transient patency
(7 to 8 months
post-infection)

3 to 4 months

Adult
heartworm
in heart
and pulmonary
vessels
(1–250 worms)

Microfilariae

L₃
14 days or longer
infective 3rd
stage larva

Adult
heartworm
in heart
and pulmonary
vessels
(1–3 worms)

Larval
development

2 to 3 months   3 to 4 days

Larval development

Larval
development

L₄
(4th stage larva)

L₄
(4th stage larva)

FIGURE 7–23 The life cycle of the heartworm

up, and swallowed into the digestive tract. In the digestive tract, the eggs hatch and larvae are passed with the feces. Larvae mature in an intermediate host such as snails or slugs, are ingested by animals, and migrate through the intestines back to the lungs (see Figure 7-24). Species affected: sheep, goats

Pinworms Small worms that travel through the large intestine and locate in the anus, where they cause itching and discomfort. Affected animals rub their tails to combat the itching. Loss of tail hair and observed frequent itching are signs that pinworms may be present. Species affected: equine

Strongyles (strohn-gīl) There are large and small varieties of strongyles. Large strongyles are the most dangerous and migrate through the body to the liver, lungs, and intestinal wall at different stages of

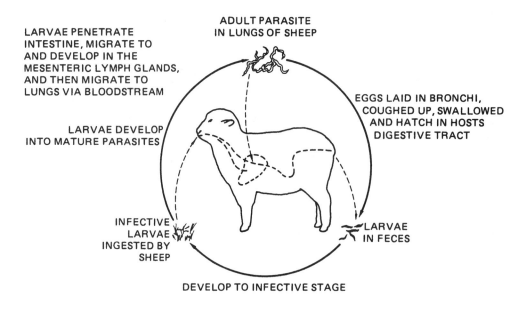

FIGURE 7–24 The life cycle of the lungworm (Courtesy of Roger Couron, Vocational Agriculture Service, University of Illinois)

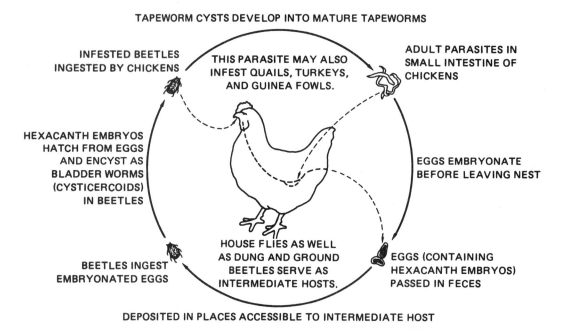

FIGURE 7–25 The life cycle of the tapeworm in poultry (Courtesy of University of Illinois at Urbana-Champaign)

the life cycle. They attach to the organs to suck blood, and cause damage at the points of attachment. They can also cause clots in the circulatory system. Small strongyles live exclusively in the intestine. Heavy loads of parasites may lead to colic in horses, and signs of malnutrition in any affected species. Species affected: canine, equine, feline

Tapeworms Segmented worms that live in the intestines of warm-blooded animals (see Figures 7-25 and 7-26). Tapeworms must spend part of their lives in an intermediate host. Animals then consume the intermediate hosts and become infected with tapeworms.

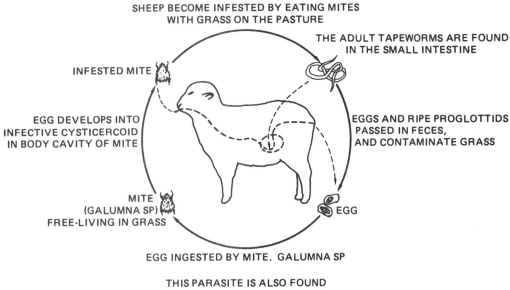

FIGURE 7–26 The life ecycle of the tapeworm in sheep (Courtesy of University of Illinois at Urbana-Champaign)

CHAPTER SUMMARY

A wide range of diseases and parasites affect animals. Diseases can be transmitted directly from animal to animal, through the environment, or through vectors. Preventing disease is an important part of animal management, and can be done through maintaining a clean and sanitary facility, appropriate vaccination, purchase of new animals that are in good health, and isolation of new animals until their health status is confirmed. Vaccination can assist animals' immune systems in fighting off disease-causing organisms, and judicious use of antibiotics can ensure the continuing good health of the animals.

STUDY QUESTIONS

1. What disease affects the central nervous system and is a zoonotic disease that affects all mammals?

2. Bacteria and viruses are examples of what type of disease-causing organisms?

3. What is the process of introducing an antigen to the body so that the immune system will respond with antibodies?

4. What is antibiotic resistance, and why is it a concern in treatment of animal disease?

5. What disease is the degeneration of a bone in the foot of the horse?

6. What ectoparasite causes mange?

7. What endoparasite is carried by mosquitoes and affects both dogs and cats?

8. What management steps can be taken to decrease the incidence of disease?

9. What is the difference between a modified live and a killed vaccine?

10. What internal parasite has the largest impact on the swine industry?

11. What is the name of the group of chemicals that are used to treat internal parasites in animals?

12. List three diseases that affect the central nervous system, and the species in which they are found.

13. What is a vector?

14. For what disease, and in what species, do many states have a law requiring vaccination?

15. Select one of the diseases listed in this chapter. Research and write a paper on the disease, including species affected, causes, symptoms, and treatment. What management practices can be used to prevent the disease?

Chapter 8
Beef Cattle

Cattle were domesticated around 6,500 B.C., and have been integral to human survival. Cattle provide meat, milk, hide, and bone, all of which have been used for food, tools, and clothing. When Europeans were colonizing different parts of the world, they brought their familiar breeds of cattle with them. In areas where native breeds existed, new breeds were often developed from crossbreeding the cattle types. Beef is a significant portion of the U.S. diet, and beef producers in the United States strive to produce a high-quality, nutritious product that will meet the evolving needs of consumers. In 2005, the United States had over 770,000 beef operations (see Figure 8-1).

BREEDS OF BEEF CATTLE

Many breeds of cattle contribute to the beef industry in the United States. Some breeds are very populous, and others fill a niche market. Most commercial animals are crossbred animals that are derived from the following breeds to maximize the qualities of each breed. Breeds are categorized by the geographical region of their origin.

American Beef Breeds

Cattle are not indigenous to the United States, and American breeds are founded on breeds from other parts of the world. The following breeds were developed in the United States from seedstock brought into America with immigration or later importation.

Beefmaster The Beefmaster was developed in Colorado in the 1930s by crossing Herefords, Shorthorns, and Brahmans. Breeders sought animals with good growth that tolerated the climactic conditions of the western United States. The breed comes in a variety of colors, and may be either horned or **polled** (pōled), which means they genetically do not have horns. Three associations register Beefmaster cattle: Beefmasters Breeders Universal, Foundation Beefmaster Association, and the National Beefmaster Association (see Figures 8-2 and 8-3).

Black baldie The black baldie is not a recognized breed, but a cross between the Hereford and the Angus. The animals are of moderate size, gain weight rapidly, and have the high-quality meat characteristic of both base breeds. They have black bodies with white faces and white markings similar to the Hereford. This cross is one of the most common in the United States, and many feedlots are full of black baldie animals.

Braford (bray-ferd) The Braford was developed in Florida by crossing Brahman and Hereford cattle. Brafords have the coloring of Herefords, with reddish bodies and white faces. The breed was developed because purebred Herefords did not thrive in the heat and humidity of Florida. Brafords grow quickly to market weight and produce high-quality meat (see Figure 8-4).

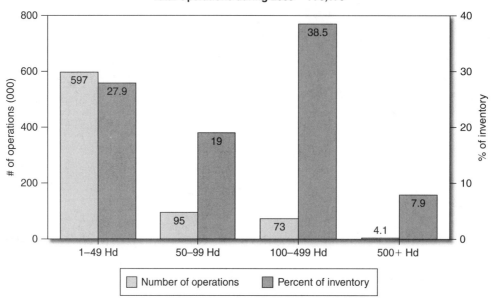

2005 U.S. Beef Cow Operations & Inventory
Total Operations during 2005— 770,170

FIGURE 8–1 Number of U.S. beef cattle operations and the inventory of cattle in 2005 (Courtesy of USDA)

FIGURE 8–2 Beefmaster bull (Courtesy of Beefmaster Breeders Universal)

FIGURE 8–3 Beefmaster cow (Courtesy of Beefmaster Breeders Universal)

FIGURE 8–4 Braford bull (Courtesy of International Braford Association, TX)

British breeds that made up the bulk of the U.S. beef industry. Brahmans range from light gray to almost black in color. The most distinctive characteristics of Brahmans are the hump on the shoulders and their large, drooping ears. Brahmans are very tolerant of heat and have a natural resistance to disease and insects. They are used primarily in crossbreeding programs with traditional beef breeds. Purebred Brahmans can be difficult to handle, and have excitable dispositions. Many bulls used in rodeo bull-riding have at least some Brahman breeding (see Figures 8-5 and 8-6).

Brahman (bra-mehn) Brahmans were developed in the southwestern United States to tolerate the dry, hot weather, by crossing **zebu** (zē-boo) cattle (*Bos indicus*) from the subcontinent of India, with the

Brangus (brayng-guhs) The Brangus was developed by crossing Brahman and Angus cattle. The breed was developed in Louisiana to combine the meat

FIGURE 8–5 Brahman bull (Courtesy of American Braham Breeders Association, TX)

FIGURE 8–7 Brangus bull (Courtesy of International Brangus Breeders Association, TX)

FIGURE 8–6 Brahman cow (Courtesy of American Brahman Breeders Association, TX)

FIGURE 8–8 Brangus cow (Courtesy of International Brangus Breeders Association, TX)

FIGURE 8–9 Santa Gertrudis cow (Courtesy of Santa Gertrudis Breeder International, TX)

quality of the Angus, with the heat and disease tolerance of the Brahman. Brangus cattle are black and polled. The **Red Brangus** is similar to the Brangus, with the exception of being red instead of black. Animals registered with either the International Brangus Breeders Association or the American Red Brangus Association must meet registration standards for conformation and size. The name *Brangus* is trademarked, and only registered animals can be identified with the Brangus name (see Figures 8-7 and 8-8).

Santa Gertrudis (sahn-tah gər-troo-dihs) The Santa Gertrudis was developed on the King Ranch in southern Texas in the early 1900s. Brahman cattle were crossed with a variety of breeds, seeking a cross that was tolerant of the climate of south Texas, and that still produced the quality of meat needed for market. The combination of Shorthorn and Brahman blood produced the bull, Monkey, who became the foundation sire of the breed. Santa Gertrudis cattle are a uniform red in color with a short, tight-hair coat. Their skin is loosely fitting, and bulls have a modified hump (see Figure 8-9).

British Beef Breeds

Many of the breeds we are most familiar with in the United States are of British origin. As a group, the cattle are of moderate size, are well-muscled, and have

manageable temperaments. This group of cattle has high-quality meat that is well marbled and highly palatable.

Angus (ayng-gəhs) Officially known as the Aberdeen Angus, Angus cattle originated in Scotland in the early 1700s. The first Angus cattle were imported into the United States in 1873. The Angus is the most populous breed in the United States, and is often used in both purebred and crossbred breeding programs. The Angus can be either black or red, and are always polled. The **Red Angus** cannot be registered in the American Angus Association, but can be registered in

FIGURE 8–10 Angus bull (Courtesy of American Angus Association, MO)

FIGURE 8–12 Galloway bull (Courtesy of Galloway Cattle Society of America, IL)

FIGURE 8–11 Angus cow (Courtesy of American Angus Association, MO)

FIGURE 8–13 Galloway cow (Courtesy of Galloway Cattle Society of America, IL)

the Red Angus Association. The red gene is recessive to the black gene, and other than coat color, the Angus and Red Angus are very similar. Both breeds produce high-quality meat that is highly palatable and well marbled (see Figures 8-10 and 8-11).

Devon (dehv-ən) One of the oldest of the British breeds, Devon cattle are a reddish color and of moderate size and muscling. They were brought to the United States in 1623 as a multipurpose animal, and were used for meat, milk, and draft. Over time, breeders have focused on the breed for its beef qualities. The breed is registered by the Devon Cattle Association, Inc.

Galloway (gahl-weigh) Galloways are a relatively small breed of cattle, but are very hardy and can tolerate harsh weather conditions. The cattle were originally bred in Scotland, and imported into the United States in 1870. The cattle have soft, relatively long and wavy hair, and come in a variety of colors (see Figures 8-12 and 8-13). The **Belted Galloway** is a popular strain characterized by black bodies with a wide white belt around the middle.

Hereford (her-ferd) The Hereford is one of the most easily recognizable of the beef breeds. The bodies of Herefords are a rich reddish color, with distinctive white faces, and white markings on the underbelly and legs. Herefords were first bred in the county of

Hereford in England. They were imported to the United States in the early 1800s, and quickly became a very popular breed. Herefords are naturally horned, with horns that grow parallel to the ground and curve back toward the face. Herefords are registered with the American Hereford Association. Polled Herefords were developed in 1910, when an Iowa breeder collected a group of naturally polled Hereford cattle, and began breeding them. Polled Herefords have the same characteristics as Herefords, with the exception of the horns. Polled Herefords can be registered either with the Polled Hereford Association, or the Hereford Association, and many animals are registered with both (see Figure 8-14).

Red Poll The Red Poll breed was developed in England by crossing native cattle from the counties of Norfolk and Suffolk. The resulting animals are solid red in color, with some white hair in the tail. Cattle were first imported into the United States in the late nineteenth century. The cattle are primarily dual purpose, although the recent focus has been on meat production. The American Red Poll Association is the registering body.

Scotch Highland The Scotch Highland was developed in the northern islands of Scotland and imported to the United States in the early 1900s, after

FIGURE 8–14 Hereford (Courtesy of American Hereford Association, MO)

FIGURE 8–15 Scotch Highland cow (Courtesy of American Highland Cattle Association, MN)

FIGURE 8–16 Shorthorn (Courtesy of American Shorthorn Association, NE)

centuries of breeding in Scotland. The animals are of moderate size and are horned, with a long coat that comes in a variety of colors. This breed is very hardy, and does well in harsh weather (see Figure 8-15).

Shorthorn The Shorthorn breed was developed in the seventeenth century in northern England, where they were originally known as Durham cattle. The breed was developed as a dual-purpose breed, for both meat and milk production, and was introduced to the United States in 1783. The cattle are red, white, or roan, or a combination of the colors, and bulls weigh up to 2,400 pounds. Although the focus of breeding in recent generations has been on developing beef traits, Shorthorns are still known as excellent milk producers and mothers. The Shorthorn has been used successfully in breeding programs, and many modern breeds have at least some Shorthorn in their backgrounds. There are both polled and horned varieties of Shorthorns, with the presence of horns being the only significant difference (see Figure 8-16).

Texas longhorn The Texas longhorn was developed from the **Spanish Andalusian** (ahn-dah-loozh-uhn) cattle that Christopher Columbus brought to the New World. The breed developed primarily in the wild state, when cattle that were taken from the Caribbean islands to Texas with the explorers escaped and adapted to the southwestern climate. The cattle are characterized by a wide variety of colors and patterns, and their distinctive horns can have a spread of more than four feet. The Texas longhorn is the breed that was most often driven on the famous cattle drives after the Civil War. Although longhorns are very tolerant of the heat and insects, the meat quality and slow maturing rate is less desirable than that of the European breeds. Longhorns nearly became extinct in the early 1900s, but are making a comeback as breeders are using them to improve the heat and disease tolerance of other breeds of cattle. The Texas Longhorn Breeding Association of America handles longhorn registration.

Continental Breeds

The Continental breeds are those that came to the United States from the Western European continent. Many of these breeds were part of the Breeds Revolution, a period of time in the late 1960s and early 1970s when many new breeds were introduced to the U.S. beef industry. These breeds were introduced in an effort to increase the size and scale of the existing breeds in the country, and to develop animals that were well suited to the feedlot system of finishing cattle.

Charolais (shar-lay) The Charolais is a French breed that was first imported to North America in 1930. The Charolais is a large breed, with mature adults weighing 2,000 pounds or more. This breed is especially heavily muscled in the loin area, and has been used extensively in crossbreeding programs to increase the size and muscling of other breeds. The American-International Charolais Association is the registering body for Charolais cattle. Charolais have **an open herdbook,** which means that crossbred animals can be registered. Refer to the rulebook requirements for precise rules (see Figures 8-17 and 8-18).

FIGURE 8–17 Charolais bull (Courtesy of American-International Charolais Association, MO)

FIGURE 8–19 Chianina bull (Courtesy of American Chiania Association, MO)

FIGURE 8–18 Charolais cow (Courtesy of American-International Charolais Association, MO)

FIGURE 8–20 Limousin steer (Courtesy of North American Limousin Foundation, CO)

Chianina (kē-ah-nē-nah) The Chianina, known casually as the **Chi** (key), was developed in Italy in ancient times. Semen was first imported to the United States in 1971. Chianina cattle are white with a black switch at the end of the tail. Chianina are a very large breed of cattle; bulls may weigh up to 4,000 pounds, and are used extensively in crossbreeding programs to increase size of the offspring (see Figure 8-19).

Gelbvieh (gehlp-fē) The Gelbvieh was developed in Germany from native cattle in the eighteenth and nineteenth centuries. They were originally developed as a multipurpose breed, but the modern Gelbvieh is a beef animal. The animals are reddish in color with no white markings, and can be either horned or polled. Gelbviehs are registered with the American Gelbvieh Association.

Limousin (lihm-ō-zēn) The Limousin is an ancient breed that was developed in France. The famous cave paintings in Lascaux depict cattle that are similar in appearance to the Limousin. Limousin cattle were first imported into the United States in 1968. Limousins are large cattle, with bulls exceeding 2,000 pounds.

Limousins are often used in crossbreeding programs to increase the leanness of the carcass, and the size of the offspring (see Figure 8-20).

Maine-Anjou (mān ahn-joo) The Maine-Anjou was developed in France as a draft animal in the 1840s. Since that time, breeding programs have focused on developing beef traits. Maine-Anjou cattle are a horned breed that is red and white in color. Mature bulls weigh over 2,500 pounds. The first Maine-Anjou crossbred calves were born in the United States in 1972 from semen imported from Canada. The American Maine-Anjou Association registers Maine-Anjou cattle (see Figure 8-21).

Marchigiana (mar-key-jah-nah) An Italian breed that was developed in ancient Rome from native cattle, including the Chianina, the Marchigiana is grayish-white, and large. Bulls weigh over 3,000 pounds. The cattle are known for their ease of calving and for being reproductively sound. Marchigianas are registered by the American International Marchigiana Society, and are used extensively in crossbreeding programs (see Figure 8-22).

Salers (sah-lair) The Salers were developed in France thousands of years ago as a dual-purpose breed. Cave paintings in southern France depict animals

FIGURE 8–21 Maine Anjou bull (Courtesy of International Maine-Anjou Association, MO)

FIGURE 8–23 Salers (Courtesy of American Salers Association, CO)

FIGURE 8–22 Marchigiana bull (Courtesy of American Milking Shorthorn Society, WI)

FIGURE 8–24 Simmental (Courtesy of American Simmental Association, MT)

believed to be ancestors of the modern Salers. Salers evolved as a very versatile breed, with uses for milk and meat, as well as serving as draft animals. Salers have been in North America since the 1970s, when they were first imported into Canada. The animals are of good size, with horns and a dark red coat color. They are used primarily in crossbreeding programs (see Figure 8-23).

Simmental (sihm-eh-tahl) The Simmental is a large breed that was developed in Switzerland in the Middle Ages. Bulls may weigh up to 2,600 pounds. The Simmental is one of the most popular breeds of beef cattle in Europe, and was imported to the United States in the late 1960s. The cattle have brown to red bodies with white faces and legs. Cows are excellent producers of milk, and Simmentals grow quickly. The American Simmental Association registers animals that meet registration requirements (see Figure 8-24).

Other Breeds

Although the majority of the beef breeds in the United States are of European origin, cattle have been successfully imported from other regions of the world. These breeds tend to fit into niche markets and specialty breeding programs, and do not constitute a large percentage of the U.S. beef industry. The following are some of those breeds:

Murray Grey Developed in Australia from crossing Shorthorn and Angus cattle, the Murray Grey semen was introduced to the United States in 1969. Although it is not a significant part of the U.S. beef industry, the Murray Grey is very popular in its native Australia. The animals are a grayish color, and are of moderate size and good temperament. The calves grow well and have good temperaments. They are used primarily for crossbreeding, and qualified animals can be registered with the American Murray Grey Association (see Figure 8-25).

Norwegian Red The Norwegian Red was developed as a dual-purpose animal in Norway. Although a minor breed in the United States, it is the most popular breed in Norway. Norwegian Reds are red, or red and white and are horned. Bulls weigh up to 2,600 pounds at maturity. Cattle are registered with the American Norwegian Red Association.

Senepol (sehn-eh-pōl) The Senepol was developed in the Caribbean by crossing the Red Poll and the N'Dama from Africa. The cattle are reddish in color, and bring many of the same characteristics to crossbreeding programs as Brahmans, with a more manageable disposition. Although the Senepol is not a populous breed, its usefulness in subtropical regions is being explored.

FIGURE 8–25 Murray Grey (Courtesy of American Murray Grey Association, MO)

CATTLE PRODUCTS

Meat is the primary product of the beef cattle industry. However, the production of beef cattle also results in the production of many marketable by-products. The beef cattle industry generates over 20 percent of the total cash receipts for agriculture. In 2004, that 20 percent equaled $40 billion dollars. Beef cattle are raised in every state in the United States, but the beef cattle industry thrives in geographical areas with adequate grazing land to maintain the cattle, or with by-products of grain production that can be fed to cattle at an economical cost (see Figure 8-26). As ruminants, cattle can consume grasses, crop residues, and other plants that are not suited for human consumption, and convert them to a highly palatable and nutritious product, beef. Often, cattle are grazed on land that cannot support crop agriculture, due to the topography, insufficient or excess water, and soil quality. The following terms relate to the processing and production of meat from beef cattle:

Aging The practice of waiting between slaughter and consumption to allow the enzymes in the meat to increase tenderness. The two primary methods of aging are **dry aging** and **wet aging.**

Dry aging The traditional way of aging beef by maintaining the meat in a temperature- and humidity-controlled environment. The meat loses moisture during the dry-aging process, which intensifies the flavor of meat. Dry aging is an expensive process because of the space and environmental management requirements.

Wet aging The meat is vacuum-packed in its own juices during the aging process. The same enzymes tenderize the meat, but the wet process does not intensify the flavor of the meat. Wet aging is less expensive than dry aging and does not require the level of environmental control required by dry aging.

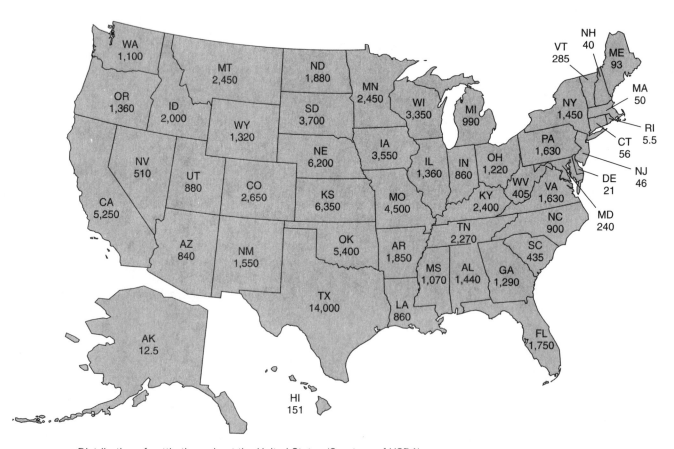

FIGURE 8–26 Distribution of cattle throughout the United States (Courtesy of USDA)

Backfat The fat that is under the skin and on the surface of the meat.

Carcass (kahr-cuhs) The body of the animal after the head, hide, and internal organs have been removed in processing.

Dark cutter An undesirable characteristic in meat that results in a darker than normal color and sticky texture. Dark cutter meat is believed to be a result of stress on the animal prior to slaughter. Although the meat is still safe to consume, it is not appealing to the consumer, and carcasses with this characteristic will be penalized in quality grading.

Grain-fed beef Those animals that have been fed primarily a grain-based diet prior to slaughter. These animals may have been raised on a primarily grass diet earlier, and then converted to a grain diet for the **finishing phase** prior to slaughter. Grain-fed beef is more tender and has an improved flavor compared to grass-fed beef, but it more expensive to produce.

Grass-fed beef Animals that are fed a forage- or grass-based diet through all aspects of growth and up to slaughter.

Hide The skin of the animal. The skin is tanned into leather, and then further processed into everything from belts and shoes to wallets and briefcases.

Marbling Intramuscular fat. Marbling is a crucial component of high-quality graded beef. More marbling results in a higher quality grade, and a more tender and flavorful cut of meat. The degrees of marbling, from most marbling to least are: **very abundant, moderately abundant, slightly abundant, moderate, modest, small, slight, traces,** and **practically devoid** (see Figure 8-27).

Quality grade A quality grade can be assessed both on live animals and on the carcass. The quality grades from highest to lowest are: **prime, choice, select, standard, commercial, utility, cutter, and canner** (see Figure 8-28).

Carcass quality grade The quality grade assessed on the actual carcass, which considers the amount of marbling in the cut and the age of the animal.

Live animal quality grade Based on the amount and quality of muscling and fat, as well as the age of the animal. Animals in the "prime" and "choice" quality grade are less than 42 months of age.

Shrinkage The amount of weight lost when transporting cattle to market.

Side of beef One half of the beef carcass.

Veal (vēl) Meat from calves less than three months of age that have been fed an exclusively milk diet. The flavor of veal is lighter than the flavor of other beef products.

Yield The weight of a chilled carcass as a percentage of the live weight of an animal. The higher the yield, the more desirable the carcass.

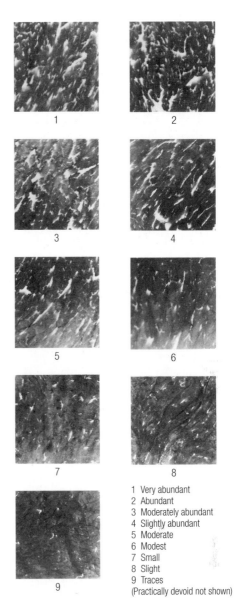

1 Very abundant
2 Abundant
3 Moderately abundant
4 Slightly abundant
5 Moderate
6 Modest
7 Small
8 Slight
9 Traces
(Practically devoid not shown)

FIGURE 8–27 Variations in marbling in beef (Courtesy of USDA. Illustrations adapted from negatives furnished by New York State College of Agriculture, Cornell University)

Yield grade A grading system for beef carcasses based on the amount of fat present. The more fat that is present, the more fat that will need to be trimmed from the finished cut, resulting in a lower yield grade. Yield grade ranges from one for the most lean, to five for the most fat (see Figure 8-29).

Retail and Wholesale Beef Products

Convenience foods A growing segment of the retail market is in foods that are ready to eat, or quickly and easily prepared. The beef industry lags behind the pork and poultry industry in production of these products, but is making progress.

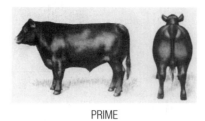

PRIME

CHOICE

SELECT

STANDARD

UTILITY

FIGURE 8–28 Quality grades of beef (Courtesy of USDA)

Ground beef   Meat that is ground and combined into a homogeneous product. Ground beef (hamburger) can be from an assortment of cuts and trimmings. If a specific cut is ground, it will be indicated in the name (for example, ground round, ground chuck, ground sirloin). Ground beef can also have a wide variation in the fat content. Because of the increased surface area of ground products, they have the highest risk of carrying pathogens. It is especially important to properly handle, store, and prepare ground meat products.

Primal cuts   Also known as wholesale cuts, primal cuts are the large segments of the carcass from which retail cuts are made (see Figure 8-30). The following are the primal cuts of beef:

Brisket primal   The front third of the underline. The major retail cut is the brisket.

Chuck primal   The front shoulder. The major retail cuts are chuck roasts, chuck steaks, shoulder roasts, and shoulder steaks.

Flank primal   The rear third of the underline. The flank steak is the retail cut.

Loin primal   The loin region. Major retail cuts are the porterhouses, T-bones, loin and tenderloin steaks, and tenderloin roasts.

Plate primal   The middle third of the underline. The skirt steak is a retail cut.

Rib primal   The thoracic vertebrae region. Major retail cuts are rib and rib eye roasts, rib and rib eye steaks, and back ribs.

Round primal   The hind leg. Major retail cuts are bottom round roasts and steaks, top round steaks, eye round roasts and steaks, round tip roasts and steaks, sirloin tip center roasts and steaks, and sirloin tip side steaks.

Shank primal   The front leg. The major retail cut is the shank cross cut.

Sirloin primal   The top of the hindquarter. Major retail cuts are the tri-tip roasts steaks, and top sirloin steaks.

Roast   A large piece of meat that can serve several people. The type of roast depends on where the meat came from on the animal. Roasting is also a method of cooking.

Steak   A smaller cut of meat than a roast. The steak often comes from the roast of the same name.

Sweetbreads   The pancreas and thyroid.

Testicles   Usually eaten fried. Also known as Rocky Mountain oysters, prairie oysters, or calf fries.

Tripe (trīp)   Stomach.

MANAGEMENT TERMS

The following are terms common in the beef industry:

Auction   Method of selling cattle by bidding in a public venue. Auctions are most often used for selling small numbers of animals.

Backgrounding   The time between weaning and going to the feedlot when a calf is fed primarily a roughage diet.

Beef checkoff   A federal program where one dollar from every beef animal sold goes to state and national organizations to promote the beef industry and fund research related to the beef industry.

Beef cycle   The fluctuation in the number of cattle owned in the United States. In approximately 10-year

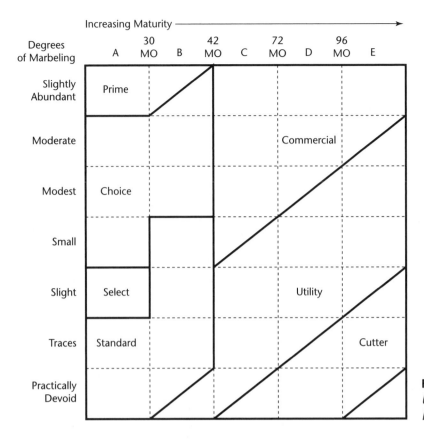

FIGURE 8–29 Yield grades of beef (Courtesy of *United States Standards for Grades of Carcass Beef,* USDA, 1997)

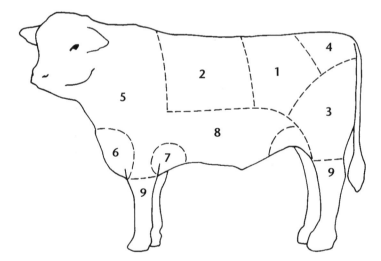

High-value wholesale cuts

1. Loin
2. Rib
3. Round
4. Rump

Low-value wholesale cuts

5. Chuck
6. Brisket
7. Flank
8. Plate or navel
9. Shank

FIGURE 8–30 Primal cuts of beef

cycles, the beef cattle numbers rise and fall due to economic pressures and the normal generation interval of cattle.

Branding The use of a tool to leave a permanent identifying mark on an animal. Most animals are branded with a hot brand, or by freeze-branding.

Breed character Characteristics unique to a breed.

Composite breed A new breed developed from combining established breeds.

Condition The amount of fat cover on a breeding animal.

Confined Animal Feeding Operations (CAFOs) CAFOs are large operations that raise animals in a limited amount of space. Facilities that are identified as CAFOs must comply with Environmental Protection Agency (EPA) guidelines regarding the disposal of waste. For details on the CAFO requirements, refer to the EPA Web site.

Corral (coh-rahl)   An area enclosed by a fence for holding animals.

Creep-feeding   The practice of providing feed to young animals in a way that adult animals cannot reach the feed. The feed may be in an area that the adult animals are too large to enter, or in a feeder that adult animals are too large to use.

Dehorning   The chemical or mechanical removal of horns.

Ear tag   An identifying tag that is inserted in the ear. This is not a permanent method of identification, as tags can be pulled out or lost (see Figure 8-31).

Ear tattoo   A permanent form of identification where marks are made on the inside of the ear.

Estrus synchronization   The practice of using hormones to have multiple cows come into estrus at the same time. This allows increased efficiency of artificial insemination and produces calves of a more uniform age.

Feed efficiency   The efficiency with which an animal turns feed into growth and meat. Increased feed efficiency results in a higher profit margin for producers, as feed is usually the most expensive input on a livestock facility.

Finish   The fat on a market animal, which can be either a steer or heifer.

Finishing phase   The final stage of preparation for market. Usually includes feeding a grain-based diet.

FIGURE 8–31 Calf with an ear tag (Courtesy of ARS)

Frame   The skeleton of the animal. Animals are judged to be large-framed or small-framed depending on the size of their skeletons.

Graze (grāz)   To eat grass.

Hazard Analysis and Critical Control Points (HAACP)   A government plant that works with packing plants to identify and resolve issues at potential areas where carcasses could be contaminated with pathogens during processing. HAACPs are also implemented on a voluntary basis in many feedlots as the beef industry strives to provide a safe food product to consumers.

Implant   A device inserted in the ear that provides a slow release of hormones to an animal. The use of an implant increases feed efficiency in animals.

Least-cost ration   A ration formulated to meet an animal's nutritional needs with the lowest cost for ingredients.

Market animal   An animal being raised for sale to a slaughter market.

Muscling   The amount of muscle throughout an animal's body.

National Animal Identification System (NAIS)   NAIS is a program of the USDA that was put in place after the September 11, 2001 attacks on the World Trade Center to develop a system to identify and track the movement of animals. According to the Animal and Plant Health Inspection Service (APHIS), implementation of NAIS will assist with "disease control and eradication, disease surveillance and monitoring, emergency response to foreign animal diseases, regionalization, global trade, livestock production efficiency, consumer concerns over food safety, and emergency management programs."

Pasture (pahs-chər)   A large enclosed area with grass for animals to eat.

Performance testing   Evaluating an individual animal based on its production in a desired area. Animals can be performance tested based on milk production, growth rate, racing speed, or any of many other characteristics.

Preconditioning   Preparation of a calf for moving to a feedlot. Preconditioning includes vaccination, castration, tattooing or branding, deworming, weaning, and treatments for internal and external parasites.

Private sale   The sale of an animal directly at a set price. Negotiation occurs between a buyer and a seller.

Range   Large tracts of land for raising animals. This term is used most often to refer to land in the Western or Plains states.

Sex character   The animal shows the characteristics of its gender. Bulls typically show more muscling and are heavier in the neck and shoulders than cows.

FIGURE 8–32 A cow in a squeeze chute (Courtesy of ARS)

Squeeze chute A form of restraint for an animal. Animals walk into chutes, where their heads are restrained in **head gates**. The animals can then be safely handled with minimal stress to the animals or the handler. Squeeze chutes are often used when doing routine health maintenance practices such as vaccination (see Figure 8-32).

Terminal markets A place where cattle are gathered and sold for slaughter. Also called a **stockyard**. Producers pay a **yardage fee** for having their animals at the stockyard until they are sold.

Tilt table Similar to a squeeze chute, the tilt table can be tilted, which is especially useful when it is necessary to treat or trim an animal's feet.

Weaning (wēn-ing) The practice of permanently separating a mother from her offspring, or removal of milk from an offspring's diet. A young animal that has been weaned is called a **weanling** (wēn-ling).

CHAPTER SUMMARY

The beef industry is a vital part of animal agriculture in the United States, and is undergoing constant change as producers strive to meet the demands of consumers. Beef is an excellent source of nutrition, and producers are using a variety of breeds and crossbreeds to raise lean animals that produce quality meat with tenderness and flavor.

STUDY QUESTIONS

Match the breed with the place of origin. Place of origin may be used more than once.

1. _____ Salers   a. France
2. _____ Hereford   b. England
3. _____ Angus   c. Scotland
4. _____ Santa Gertrudis   d. Germany
5. _____ Limousin   e. India
6. _____ Chianina   f. United States
7. _____ Zebu   g. Australia
8. _____ Beefmaster   h. Switzerland
9. _____ Simmental   i. Italy
10. _____ Murray Grey   j. Texas

11. Which of the following breeds is used extensively in crossbreeding to increase heat and disease tolerance?
    a. Hereford
    b. Brangus
    c. Brahman
    d. Simmental

12. What beef breed also has a dairy breed associated with it?
    a. Hereford
    b. Shorthorn
    c. Murray Grey
    d. Angus

13. The following crossbred is a result of breeding Angus and Herefords:
    a. Beefmaster
    b. Brangus
    c. Simmental
    d. Black baldie

14. What term describes an animal's skeletal size?
    a. Frame
    b. Stature
    c. Mass
    d. Conformation

15. _____ is the method of feeding young calves that prevents older animals from accessing the feed.

16. List three beef breeds that are composite breeds, and the breeds that were combined to develop them.

17. List the primal, or wholesale, cuts of beef.

18. From what primal cut do we get T-bone and porterhouse steaks?

19. What is the current price per pound that producers are receiving for beef? Compare to the current cost per pound for beef in the grocery store.

Chapter 9
Dairy Cattle

Chapter Objectives

▶ Learn the primary breeds and origins of dairy cattle

▶ Learn the states that are primarily involved in the dairy industry

▶ Be familiar with the primary products of the dairy industry

Cattle have been dual-purpose animals through much of human history. Farmers owned cattle that produced both meat and milk for market, and were often also used as draft animals. In the mid-nineteenth century, breeders began focusing on developing the milk and meat production characteristics of their cattle. Dairy production is part of animal agriculture in all 50 states, but California has the most cows, and produces the most pounds of fluid milk. Long known as "America's Dairyland," Wisconsin was the leading state in fluid milk production until the 1990s, and is still the leading state in regard to production of manufactured milk products, such as cheese and butter. However, the gap is closing, and California may surpass Wisconsin in the production of manufactured milk products soon. In general, dairies are numerous in states with large populations due to the need to get fluid milk processed and to the stores as quickly as possible. Following California and Wisconsin in annual milk production (2006) are New York, Pennsylvania, and Idaho. The dairy industry comprises approximately 11 percent of the total income generated by agriculture, for a value of approximately $22 billion in 2004, according to the United States Department of Agriculture. Most dairy operations (almost 80 percent) have fewer than 100 cows, and many are still family-owned businesses (see Figure 9-1).

BREEDS OF DAIRY CATTLE

The dairy industry in the United States is focused on only a few breeds of cattle. Although many commercial dairy cattle are not registered, most are purebred. Each breed has different strengths and differences in the components of the fluid milk produced. The end product of the milk is one of the determinants in choosing the breed to milk. Breeds that produce milk higher in butterfat and proteins may be selected for producing milk sold for processing, whereas producers selling fluid milk choose breeds that produce a higher volume of milk. Table 9-1 shows the differences in the volume of milk produced annually, and milk components for the major breeds. Dairy breeds are often named for the country, or region, or origin. The following is a list of the major breeds used in the United States dairy industry.

British

The following are British breeds of dairy cattle:

Ayrshire (air-shīr) This breed was developed in Scotland in the late 1700s and was imported into the United States in the 1800s. Ayrshires are red and white, with the shade of red ranging from brown to mahogany. They should have sharp lines between the red and white patches on the hide. Roaning, the mixing of red and white hairs, is undesirable.

113

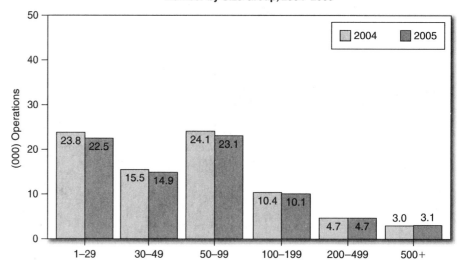

U.S. Milk Cow Operations
Number by Size Group, 2004–2005

FIGURE 9–1 Number of dairy operations in the United States (Courtesy of USDA)

TABLE 9–1

Production of milk and milk components by different cattle breeds in the United States

2006 Herd Averages of Cows on DHI[1] testing						
Breed	**Number of Herds**	**Milk (lbs/cow)**	**BF[2] % (%)**	**BF # (lb)**	**Protein % (%)**	**Protein # (lb)**
Ayrshire	115	15,515	3.91	606	3.16	490
Brown Swiss	275	18,152	4.06	742	3.37	617
Guernsey	170	15,402	4.52	697	3.35	517
Holstein	21,376	22,569	3.66	829	3.06	693
Jersey	1,154	16,127	4.61	744	3.59	579
Milking Shorthorn	41	14,676	3.73	547	3.10	454
Red and White Holstein	29	20,395	3.69	752	3.04	620
Mixed	1,376	17,829	3.96	712	3.22	580
All breeds	24,536	22,061	3.70	820	3.08	682

Source: DHI Report K-3. http://aipl.arsusda.gov/publish/dhi/current/hax.html.
[1]Dairy Herd Improvement
[2]Butterfat

Animals are of moderate size, and should reach approximately 1,200 pounds at maturity. Ayrshires used to be known for their distinctive horns, which curved up and toward the back; however, now virtually all animals are dehorned. The cattle were developed to produce efficiently in the sometimes-harsh Scottish climate, and modern Ayrshires still have an ability to forage well. Ayrshires are also known for having exceptionally sound mammary glands, and feet and legs. Ayrshires are registered with the American Ayrshire Association (see Figure 9-2)

FIGURE 9–2 Ayrshire cow (Courtesy of Ayrshire Breeders' Association, VT)

FIGURE 9–3 Guernsey cow (Courtesy of American Guernsey Cattle Club)

FIGURE 9–4 Jersey cow (Courtesy of American Jersey Cattle Club)

Guernsey (gurn-zē)   The Guernsey was developed on the Isle of Guernsey in the English Channel from foundation stock of French cattle around A.D. 960. Guernseys were introduced into the United States in the 1830s, and are moderate producers of milk with a unique golden color. Guernseys are a medium-framed breed, with a golden and white coat color. Guernseys are second only to Jerseys in the amount of fat produced in the milk, and are popular with people wishing to make cheese and butter from milk. The American Guernsey Association registers Guernsey cattle (see Figure 9-3).

Jersey (jər-zē)   The Jersey was developed on the Isle of Jersey in the English Channel. The precise foundation stock is unclear, but in 1763, law forbade the importation of cattle to the isle, resulting in the development of this breed. The Jersey is the smallest of the dairy breeds, with a mature weight of around 900 pounds. They are a light fawn to black color with a distinctively dished face and large dark eyes. Jerseys are very attractive cattle with excellent udder conformation and dairy quality. They also use feed very efficiently. They produce the highest volume of milk per pound of animal weight of any of the dairy breeds. Jerseys are fifth of the six primary breeds in regard to total volume of milk produced, but they are significantly higher in butterfat production than the other breeds, which is important for butter and cheese production. Jerseys have more excitable temperaments than the other dairy breeds, and this is especially true of Jersey bulls, which have earned a reputation for being difficult to handle for even the most experienced dairymen. Jerseys can be registered with the American Jersey Cattle Club (see Figure 9-4).

Milking Shorthorn   The Shorthorn breed was developed in the seventeenth century in northern England, where they were originally known as Durham cattle. The breed was developed as a dual-purpose breed, for both meat and milk production, and was introduced to the United States in 1783. The cattle are red, white, or roan, or a combination of the colors, and bulls weigh up to 2,400 pounds. The foundation Shorthorn was developed into two breeds: the beef qualities were emphasized to develop the Shorthorn, and the milk-producing qualities were emphasized to produce the Milking Shorthorn breed. Milking Shorthorns can be registered by the American Milking Shorthorn Society or in the American Shorthorn Association.

Continental Breeds

The following are continental breeds of dairy cattle:

Brown Swiss   The Brown Swiss was developed in Switzerland, and is believed to be one of the oldest breeds of dairy cattle. Brown Swiss cattle were initially bred as dual-purpose animals, and their size and strength made them popular draft animals. The cattle are light to dark brown, slow to mature, long-lived, and produce a volume of milk second only to the Holstein. Brown Swiss were introduced to the United States in the late 1800s, but the majority of cattle in the country are a result of U.S. breeding programs, not importation. Brown Swiss can be registered with the Brown Swiss Cattle Breeders' Association of the USA (see Figure 9-5).

FIGURE 9–5 Brown Swiss cow (Courtesy of Brown Swiss Cattle Breeders' Association of the USA)

116 *Chapter 9*

FIGURE 9–6 Holstein-Friesian cow (Courtesy of ARS)

Holstein-Friesian (hōl-stēn frē-zhəhn) The Holstein-Friesian (commonly known as Holstein) was developed in the Netherlands and is the most popular dairy breed in the United States, with more than 90 percent of all dairy cattle in the United States being of this breed. Holsteins have a distinctive black-and-white color pattern that can range from predominately black with white markings, to predominately white with black markings. Holsteins do possess a recessive gene for the red color, and red Holsteins are occasionally seen. Holsteins produce the highest volume of fluid milk per cow of any breed, although they do have the lowest butterfat production of the common dairy breeds. Holsteins are the largest of the dairy breeds, with cows reaching a mature weight of 1,500 pounds. Calves are around 90 pounds at birth. Cattle can be registered with the Holstein Association USA (see Figure 9-6).

DAIRY PRODUCTS

Dairy cattle produce a wide range of products for human consumption. Fluid milk consumption has decreased in recent years (see Table 9-2), for many reasons. In addition to increased competition in the beverage market, an increasing percentage of the U.S. population is comprised of ethnic groups who have a higher incidence of **lactose intolerance.** People that are lactose intolerant cannot properly digest fluid milk, although often consume products such as yogurt to gain the health benefits of dairy product consumption. In addition to the milk-related products produced by dairy cattle, dairy cattle are also used for meat production. Dairy animals that are not selected for the milking herd are raised to market weight and sold in the same manner as beef cattle. In addition, dairy cattle that have completed their milk production career also are sold as meat animals. Holsteins are especially suited as "dairy beef" because of their large frame size. The following are additional products produced by the dairy industry:

Butter The product made from the milkfat portion of the milk. Butter is made by agitating the milk or cream until the fat particles bind together.

Buttermilk The fluid left after butter has been removed from the milk.

Cheese A product made from whole milk with the addition of enzymes to coagulate the solids into cheese. There are hundreds of varieties of cheese, and many regions and countries have cheeses unique to their areas. Cheese consumption per capita has increased in the United States over the last five years (see Table 9-3). Cheese can be made from any type of milk, including sheep and goat milk.

TABLE 9–2
U.S. Per capita beverage milk availability

	Gallons per capita per year		
Year	Total plain and flavored whole milk	Total lower-fat and skim milk	Total beverage milk
2005	6.9	14.0	21.0
2004	7.3	13.9	21.2
2003	7.6	13.9	21.6
2002	7.7	14.2	21.9
2001	7.8	14.2	22
2000	8.1	14.4	22.5

Low-fat and fat-free milk include 2% reduced-fat milk, low-fat milk (1%, 0.5%, and buttermilk), and skim milk (fat-free). Calculated from unrounded data.

Source: USDA/Economic Research Service. Last updated Feb. 15, 2007.

TABLE 9–3

U.S. per capita cheese product availability

| | Pounds per capita per year | | |
Year	American cheese	Other cheese	Total cheese
2005	12.7	18.7	31.4
2004	12.9	18.3	31.3
2003	12.5	17.9	30.4
2002	12.8	17.6	30.5
2001	12.8	17.2	30
2000	12.7	17.1	29.8

Natural equivalent of cheese and cheese products. Excludes full-skim American and cottage, pot, and baker's cheese. American cheese includes cheddar, Colby, washed curd, stirred curd, and Monterey Jack. Other cheese includes Romano, Parmesan, mozzarella, ricotta, other Italian cheeses, Swiss, brick, Muenster, cream and Neufchatel, blue, Gorgonzola, Edam, Gouda, imports of Gruyere and Emmentaler, and other miscellaneous cheeses. Calculated from unrounded data.

Source: USDA/Economic Research Service. Last updated Feb. 15, 2007.

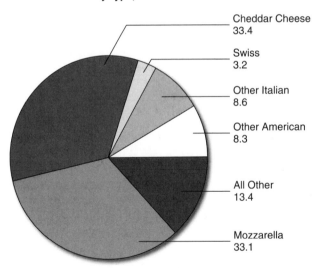

Cheese Production Percent by Type, 2005

Cheddar Cheese 33.4
Swiss 3.2
Other Italian 8.6
Other American 8.3
All Other 13.4
Mozzarella 33.1

FIGURE 9–7 Production of different types of cheese in the United States. (Courtesy of USDA)

Cheeses can be aged or fresh. Mozzarella and cheddar are the two most popular cheeses produced in the United States (see Figure 9-7).

Condensed milk A canned milk that has had 50 percent of the water removed. Also known as evaporated milk. Sweetened condensed milk has added sugar.

Curd (kerd) The solid that is created in the cheese-making process after the addition of enzymes.

Cream The portion of milk containing milkfat. In raw milk, cream rises to the top of the milk when it is cooled. Commercially available cream is divided into categories based on the amount of milkfat. Light cream is 30–38 percent milkfat, heavy cream has more than 48 percent milkfat.

Dairy beef Dairy cattle that are sold for meat. Most veal is from dairy calves, many dairy steers are fed in feedlots and sold like beef cattle, and most culled dairy cattle are sold for meat.

Evaporated milk Milk that has had 50 percent of the water removed, and is then canned. This is the same as unsweetened condensed milk.

Fluid milk Milk sold for consumption in its liquid form. Fluid milk is categorized in the store by the percentage of fat in the milk, from skim and nonfat, to whole milk, which has had no milkfat removed. California is the top producer of fluid milk in the United States. Holstein cattle produce significantly more fluid milk than any other breed. Both annual milk production per cow, and total annual milk production in the dairy industry have increased over the last 10 years (see Figures 9-8 and 9-9).

Half-and-half A fluid product that is 50 percent cream and 50 percent whole milk.

Homogenized milk Milk that has been processed so that the milkfat is distributed throughout the milk and does not separate.

Ice cream A frozen dairy product containing at least 10 percent milkfat. Premium ice creams have higher milkfat than economy brands. **Ice milk** is similar to ice cream, but has milkfat of less than 10 percent. Table 9-4 shows the consumption of ice cream and frozen dairy products. More than 950 million gallons of ice cream were produced in the United States in 2005.

Pasteurization (pahs-tər-ī-zā-shuhn) The process of heating milk to kill microorganisms that cause spoilage or health risks. All commercially available milk is pasteurized.

Raw milk Milk that is fresh from the cow and has not been pasteurized or homogenized.

Rennet (rehn-it) The product containing the enzyme **rennin** (rehn-ihn) that is added to milk to coagulate the milk and form curds.

Whey (way) The liquid left after cheese has been made. Dried whey is often used as a feed product and is a good source of bioavailable protein. An increasing market for whey is in the form of human protein supplements, and its use in animal feed may decrease as its value increases.

Yogurt (yō-gərt) A dairy food made from the bacterial fermentation of milk.

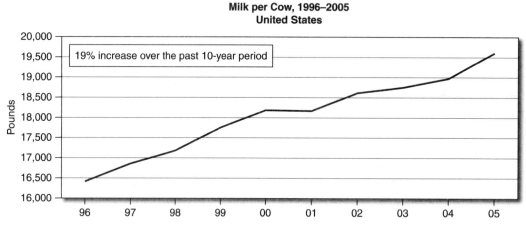

FIGURE 9–8 Annual milk production per cow. (Courtesy of USDA)

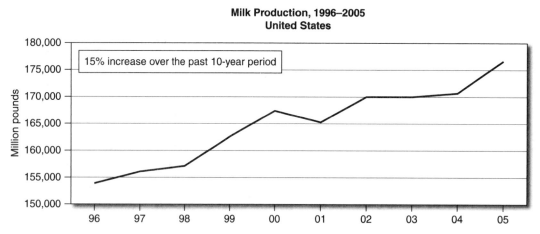

FIGURE 9–9 Total milk production in the U.S. (Courtesy of USDA)

TABLE 9–4

U.S. per capita frozen dairy products availability

	Pounds per capita per year					
Year	Ice cream	Low-fat ice cream	Sherbet	Frozen yogurt	Other frozen	Total frozen dairy
2005	15.4	5.9	.89	1.3	.57	24.1
2004	15.0	7.2	1.1	1.3	0.63	25.3
2003	16.4	7.5	1.2	1.4	0.57	27.1
2002	16.7	6.5	1.3	1.5	0.62	26.6
2001	16.3	7.3	1.2	1.5	0.69	27
2000	16.7	7.3	1.2	2	0.9	28

Low-fat ice cream is formerly known as ice milk and includes small amounts of nonfat ice cream. Other frozen products include nonstandardized frozen dairy products not listed separately. Calculated from unrounded data.

Source: USDA/Economic Research Service. Last updated Feb. 15, 2007.

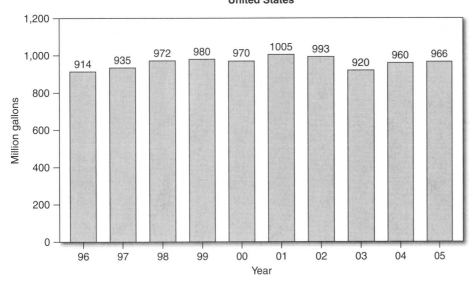

Regular Ice Cream, U.S.
United States

(Chart: Million gallons by Year)

Year	Million gallons
96	914
97	935
98	972
99	980
00	970
01	1005
02	993
03	920
04	960
05	966

FIGURE 9–10 Ice cream produciton in the United States. (Courtesy of USDA)

DAIRY MANAGEMENT TERMS

The following are common terms associated with dairy management:

Babcock Cream Test A method of determining how much milkfat is in a milk sample. The amount of milkfat is involved in determining the price for milk.

Bulk tank A tank on the dairy farm that is used to store collected cow milk until it is picked up by the milk cooperative. The bulk tank cools the milk to 40 degrees, and maintains that temperature until transport arrives (see Figure 9-11).

Bovine somatotropin (BST) (soh-mah-tah-trō-pihn) BST is a naturally occurring hormone that the cattle secrete from the pituitary gland. Recombinant BST (rBST) can be commercially manufactured and is available for producers to use for the purpose of increasing milk production.

Bull stud A facility that houses bulls that are used in artificial insemination breeding programs. Most dairy cattle are bred via artificial insemination with frozen semen from bulls housed at bull studs.

Butterfat Same as milkfat.

California Mastitis Test The California Mastitis Test can be done instantly at the side of a cow to determine the presence of mastitis. If the cow has mastitis, the milk cannot be introduced to the bulk tank or sold.

Calf hutch A small shelter, similar in appearance to a doghouse, for housing calves. Hutches have a small fenced area so calves can go in and out of the shelter. Disease spreads less with the use of calf hutches rather than group housing situations (see Figure 9-12).

Calving interval The amount of time between calves. Twelve to thirteen months between calves is considered ideal.

Casein (kā-sēn) The primary protein present in milk.

Cleanup bull A bull that may be turned out to impregnate cows that were not artificially inseminated.

Cow trainer An electrical device suspended over the back of cattle in tie or stanchion stalls. If a cow is too far forward to reach the gutter when arching her back to defecate or urinate, the device is set so that she will receive a mild shock until she backs up to reach the gutter.

Culling The permanent removal of an animal from a herd. Cattle are culled for low production, health, or reproductive problems.

Dairy character A term used to describe how closely an animal fits a dairy type. Animals with dairy

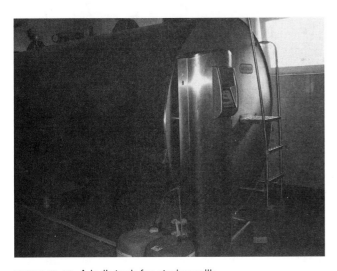

FIGURE 9–11 A bulk tank for storing milk.

FIGURE 9–12 A calf housed in a calf hutch. (Photo by Michael Dzaman)

FIGURE 9–13 A free stall housing situatin for dairy cattle (Courtesy of Mike Schutz, Purdue University)

FIGURE 9–14 A gutter and gutter cleaner. (Photo by Michael Dzaman)

character are lean and do not have excessive fat cover. Dairy animals should focus their energy on production of milk, not meat and fat, and dairy character is a visual representation of that distribution of energy.

Dairy Herd Improvement Association (DHIA) An organization that collects and maintains production information on cattle in member herds. Information is then returned to producers to enable them to make management decisions.

Dehorning The mechanical or chemical removal of horns. Horns on dairy cattle serve no purpose, and cattle with horns can injure each other and people. It is less stressful to remove the horns on young calves than on older animals.

Dry cow A cow that is not currently producing milk. Cattle are typically "dried off" 45–60 days prior to delivery of a new calf.

Free-stall barn A large housing system with stall areas and feeding areas. Cattle freely move throughout the housing system and are not confined to the stalls (see Figure 9-13).

Freshen (frehsh-ehn) To deliver a calf and return to milk production, thereby beginning a "fresh" production cycle.

Grade An animal that is not registered with a breed association. The animal can either be a purebred (a grade Holstein), or of an unknown breed (a grade cow).

Gutter The channel behind cows in stanchions that collects manure that is deposited during the milking process. The gutter cleaner is the mechanical device that moves the manure out of the barn to the manure collection area (see Figure 9-14).

Herd health program A health care plan that maintains the health of a herd of animals. A herd health program includes all aspects of herd management: vaccination, nutrition, reproduction, and facility management.

Hoard's Dairyman The magazine published in Wisconsin since 1885 that is the premier dairy magazine in the United States.

Lactation (lahck-tā-shuhn) The production and secretion of milk.

Lactation curve   The plot of the volume of milk produced over the length of a lactation period. Volume peaks approximately 60 days into the lactation period, and gradually declines over time. *Persistence* is the term used to describe how long production stays at a high level before it begins to decline. The more time dairy cattle spend in the early phase of the lactation curve, the more milk they will produce over their lifetimes. Therefore, dairy producers strive to have cows produce a calf every 13 months.

Linear classification   A method of evaluating dairy cattle that assigns a standardized scoring system to different heritable traits.

Milk check   The paycheck farmers receive for their milk. The check is the price of the milk per hundredweight, minus the cost of hauling the milk.

Milkfat   The fat portion of the milk. Different breeds show significant differences in milkfat production, with Jerseys having the highest milkfat production at over 5 percent. Milkfat is used for production of cheese and butter (see Table 9-1).

Milk grades   Farms are certified to produce grades of milk based on standards of facility maintenance and herd health. The following are grades of milk:

Grade A milk   The highest-quality milk, and the only milk that can be sold in the United States as fluid milk. Approximately 85–90 percent of milk produced in the United States meets Grade A standards.

Grade B milk   Milk produced at facilities that do not meet the standards for Grade A milk, Grade B milk is used for further processing such as butter or cheese. The further processing ensures safety for consumption.

Milk house   The place that houses the bulk tank and all the equipment for collecting, cooling, and storing milk.

Milk letdown   The release of milk from the alveoli in the udder to allow removal by the milking machine or calf. Cattle can be conditioned to let down milk when they enter the milking environment, or have their udders washed.

Milking machine (milker)   A machine that uses a vacuum and an inflation tube and teat cup assembly to remove milk from cows. The following are the three primary types of milking machines:

Pail milkers   Milk is removed from cows into containers that sit next to the cows. Pails are then carried to the bulk tank and emptied by hand.

Pipeline milkers   Milk is removed from cows and follows a pipeline through the barn to the bulk tank. The person milking does not handle the milk containers (see Figure 9-15).

Suspension milkers   Similar to pail milkers, except they are suspended off the ground by a **surcingle** (ser-sing-gel) around the cow's body.

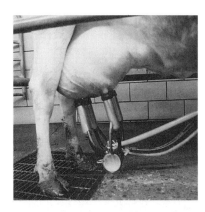

FIGURE 9–15 A cow hooked to a pipeline milking machine. (Photo by Michael Dzaman)

FIGURE 9–16 Jersey cattle in a milking parlor (Courtesy of Mike Schutz, Purdue University)

Milking parlor   A system for milking in which cattle enter in groups, and a person in a recessed pit attaches the milking machine. Parlors can be arranged in herringbone, side-opening, carousel, or polygon designs (see Figure 9-16).

Production records   Records that show how much milk and butterfat each cow has produced. The record also provides information on age, frequency of milking, and any health-related information.

Purebred   An animal that contains genes of only one breed. Purebreds can be registered or unregistered.

Replacement heifers   Young females maintained to enter a milking herd.

Registered   An animal whose parentage is documented with an established association.

Stanchion (stan-chun)   A head gate that restrains cows for milking. Barns with stanchions usually have all the cows come in at one time, and stand in rows, to be milked. Cows may come in only for milking, or may spend the majority of their time in the barn. If

FIGURE 9–17 Cattle in a stanchion barn (Photo by Michael Dzaman)

FIGURE 9–18 Holstein cattle in a tie stall barn. Note the Red Holstein in the foreground. (Courtesy of Jeff Bewley, Purdue University)

they live primarily in the barn, mats or other bedding are provided so the animal can lie down comfortably (see Figure 9-17).

Strip cup A cup used to hold a small amount of milk that is manually milked out before the machine is attached. Milking a bit into the strip cup encourages milk letdown and reduces bacteria in the milk.

Teat dip A disinfectant that teats are dipped in after milking to prevent bacteria from entering the mammary glands.

Teat removal Some heifers are born with extra teats. These teats should be removed at a few weeks of age to prevent problems with infection in the future.

Tie stall Similar to stanchions in that each cow is restrained in its place in the barn. Cows are restrained by a neck chain or strap (see Figure 9-18).

Total mixed ration (TMR) A method of combining ration ingredients so each bite contains all necessary nutrients for the animal.

Type classification A method of classifying animals based on their physical characteristics, ranging from excellent to poor. A registry representative for the breed assigns the classification, and gives a numerical score to each animal.

CHAPTER SUMMARY

The dairy cattle industry is very different from the beef industry. Whereas the beef industry has a wide variety of breeds, and use of crossbred animals is extremely common, more than 90 percent of the dairy cattle in the United States are of the Holstein breed, and only six breeds are represented in the U.S. dairy industry. There is a current trend toward using some crossbreeds in herds, especially crosses of Jerseys or Brown Swiss on Holsteins. Some programs are also experimenting with the introduction of European breeds to try to increase fertility rates. However, at this time, the overwhelming majority of U.S. dairy farms continue to milk exclusively Holstein cattle.

Dairy operations require a large amount of specialized equipment for the safe handling of milk, and cattle must be milked at least twice a day to maximize production. Most of the dairy operations in the United States are under 200 cows, and are family-owned and operated. However, a growing number of extremely large dairies are corporately owned. The dairy industry has developed standardized methods for evaluating the physical appearance and production of cows, and producers use that information to make management decisions about the herd.

STUDY QUESTIONS

Match the dairy breed with its characteristic:

1. _____ Jersey   a. Produces the largest volume of milk.

2. _____ Ayrshire   b. Golden brown and white in color.

3. _____ Brown Swiss   c. Red and white with roaning acceptable.

4. _____ Holstein   d. Produces the highest percentage of butterfat.

5. _____ Milking Shorthorn   e. Originally from Scotland.

6. _____ Guernsey   f. Bred as a dual-purpose milking and draft animal.

7. What state has the most dairy cattle?
   a. Wisconsin
   b. Texas
   c. California
   d. New York

8. What state produces the most cheese and butter?
   a. Wisconsin
   b. Texas
   c. Idaho
   d. New York

9. Which is used as a house for young calves?
   a. Parlor
   b. Tie stall
   c. Free stall
   d. Gutter

10. What is the DHIA, and what does it do?

11. What is the purpose of teat dip?

12. List three dairy products other than fluid milk.

13. What is the current price that producers receive for fluid milk in your state?

14. What is the difference between ice milk and ice cream?

15. Obtain a container of cream from the grocery store. Pour the cream into a jar with a tight lid. Make sure the jar is large enough that the cream fills it between half and three-quarters full. Shake the cream until it divides into a solid and liquid portion. What is each portion? If one person uses heavy cream and another uses light cream, which will produce more of the solid product? Why?

Chapter 10
Swine

Chapter Objectives

▶ Learn the basic structure of the swine industry

▶ Learn the primary breeds in the U.S. swine industry

▶ Learn the wholesale and retail cuts of pork

▶ Be familiar with common swine management terms

Hogs were domesticated in 6,500 B.C., and have been an important part of human history. Hogs are very efficient at converting feedstuffs to a palatable meat product. Because hogs are omnivores and will eat and utilize both plant-based and animal-based feeds, they are an important part of agriculture throughout the world. In the United States, where producers have access to high-quality grain, most hogs are fed a grain diet; in developing countries that may have little or no grain production beyond the needs of the human population, hogs can convert a wide variety of feedstuffs into meat. Hogs are excellent foragers, and although production on a foraging diet is less efficient than on a grain-based diet, they are an excellent livestock choice in many developing countries.

When the European explorers came to the New World, they brought domestic hogs with them. Until the 1950s, fat was a major product, and hogs were bred and raised for the fat (**lard**) as much as for meat production. However, modern hog producers focus on breeding and raising animals that produce lean, nutritious meat. Due to its ability to be responsive to consumer desires, the swine industry continues to be strong in the United States. As consumer demand shifted toward leaner products, swine producers met that need, resulting in a consistent demand for pork over the years (see Figure 10-1). Swine populations are greatest in areas where ample grain is available to feed the animals with minimal transportation cost for the feed. As in the beef industry, the majority of market animals are crossbred animals.

BREEDS

Although most hogs produced for meat in the United States are crossbred animals, it is important to understand the traits and characteristics of each of the breeds that are combined to create these crossbreeds. Crossbreeds are developed to combine the traits of both of the contributing breeds. Some crosses are bred with the focus on producing "maternal traits" that contribute to success and efficiency in raising offspring. Other crosses are bred with the focus on "production traits" that are those that contribute to the production of a high-quality, uniform, and marketable carcass. Uniformity of the carcass is very important in the U.S. pork industry. Processing is highly mechanized, and carcasses of uniform size and shape are processed at a greater speed, and with a greater amount of marketable meat from each carcass. In countries where animals are less uniform coming to slaughter, a greater amount of the work must be done by hand to maximize the amount of usable meat from each carcass.

American Breeds

Chester White The Chester White breed was developed in the mid-nineteenth century in Pennsylvania by crossing Yorkshire, Lincolnshire, and Cheshire

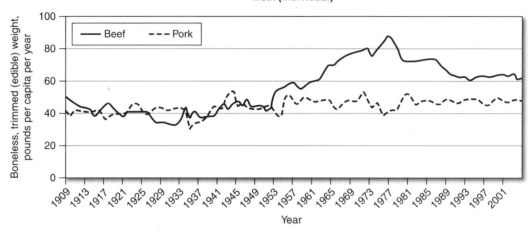

U.S. per Capita Food Consumption
Meat (individual)

Figures are calculated on the basis of raw and edible meat. Excludes edible offals, bones, and viscera for red meat. Excludes game consumption for red meat. Calculated from unrounded data.

Source: USDA/Economic Research Service. Last updated Dec. 21, 2005.

FIGURE 10–1 Comparison of pork consumed and beef consumed over the years (Courtesy of USDA)

FIGURE 10–2 Chester White boar (Courtesy of Swine Genetics)

FIGURE 10–3 Chester White gilt (Courtesy of Chester White Swine Record, IL)

FIGURE 10–4 Duroc gilt (Courtesy of United Duroc Swine Registry, IL)

the exact breeds involved in the development of this distinctive red hog are unclear, Berkshires are probably an ancestor, and there may also be breeding from hogs imported from Africa. The Duroc is a popular breed in the United States. Its color ranges from a light yellow to a dark mahogany red. The ears flop slightly forward and the hogs typically grow quickly. Durocs can be registered with the United Duroc Swine Registry (see Figure 10-4).

Hereford (her-fərd) The Hereford breed was developed in Iowa, Missouri, and Nebraska from Poland China, Duroc, Chester White, and Hampshire hogs. The hogs are similarly marked to Hereford cattle, with a red body, white face, and white feet. Marking requirements are strict and, according to the National Hereford Hog Registry Association, must ". . . have a white face, not less than two-thirds red exclusive of face and ears, with at least two white feet—white showing not less than one inch above the hoof. They must be red in color, either light or dark red." Animals not meeting these requirements cannot be exhibited or bred as Hereford hogs. In addition to their unique

hogs. Those hogs were then crossed with a boar from England that was either from Cumberland or Bedfordshire. The Chester White is a moderate-sized white hog with forward flopping ears. Chester Whites are registered with the Chester White Swine Record (see Figures 10-2 and 10-3).

Duroc (doo-rok) The Duroc was developed in the eastern United States in the mid-1800s. Although

coloring, Herefords are known as good mothers that birth similar numbers of piglets to other major hog breeds. Hereford hogs are registered with the National Hereford Hog Registry Association.

Poland China   The Poland China hogs are not from Poland or China, but originated in Ohio in the early 1800s from crossing Russian, Byfield, Big China, Berkshire, and Irish Grazer stock. Poland China hogs are large black hogs with white feet, faces, and tails. They have forward flopping ears, and are known for having low **backfat** and large **loin eyes**. Poland Chinas are registered with the Poland China Record Association (see Figure 10-5).

Spotted Swine   Spotted Swine are commonly known as Spots Hogs, and were developed in Indiana from crossing Poland China hogs with native hogs. Gloucester Old Spots were later added. The animals are large-framed, like the Poland China breed, and the body is covered with black and white spots. In the 1970s, breeders were allowed to reintroduce new

Poland China blood to the breed for a few years. Eligible animals can be registered with the National Spotted Swine Record (see Figures 10-6 and 10-7).

British Breeds

Berkshire (bərk-shər)   The Berkshire breed was developed in the early eighteenth century in the Berkshire region of England. The hogs are black with white feet, faces, and tails. Berkshire breeders have been dedicated to maintaining the purity of the breed, while still judiciously using the animals in crossbreeding programs and for the development of other breeds. The American Berkshire Association was founded in Illinois in 1875, and was the first registry for purebred swine in the world. This organization registers Berkshires (see Figure 10-8).

Hampshire (hahmp-shər)   The black and white Hampshire breed was developed in the Hampshire region of England. Its primary physical characteristic is the distinctive white belt around its front quarter that includes the front legs. The Hampshires have erect ears, and are known for having lean carcasses, which are very desirable in the modern swine industry. Hampshires are commonly used in crossbreeding programs. Purebred animals can be registered with the Hampshire Swine Registry (see Figure 10-9).

Tamworth (tahm-wərth)   Tamworths are red hogs with erect ears and long faces. They originated in Ireland and England in the early nineteenth century. The Tamworth breed is one of the oldest pure breeds

FIGURE 10–5   Poland China boar (Courtesy of Poland China Record Association, IL)

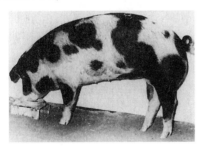

FIGURE 10–6   Spotted gilt (Courtesy of National Spotted Swine Record, Inc., IL)

FIGURE 10–7   Spotted barrow (Courtesy of National Spotted Swine Record, Inc., IL)

FIGURE 10–8   Berkshire boar (Courtesy of American Berkshire Association, IN)

FIGURE 10–9   Hampshire boar (Courtesy of Hampshire Swine Registry, IN)

of hogs. They have large litters, and females have good mothering abilities. They are especially known as producers of high-quality bacon. Tamworths are also very good foraging hogs, and do well in subsistence and home-farming situations where they have the opportunity to forage. Tamworths can be registered with the Tamworth Swine Association.

European Breeds

Landrace (lahn-drās)   The Landrace breed is a large white hog that originated in Denmark. Landraces are known for producing large litters of piglets (see Figure 10-10), and for being good mothers that raise most of the piglets to weaning age. Landraces are characterized by relatively long, lean bodies, and ears that flop down over their eyes. Landraces are registered with the American Landrace Association (see Figure 10-11).

Pietrain (pē-ah-trān)   The Pietrains were developed in Belgium in the 1950s. Pietrains are medium-sized with white and black spots and erect ears. Pietrains have extremely muscled hams. They are used primarily in crossbreeding programs to take advantage of their exceptional leanness of carcass. They do lack in mothering ability and milk production, so boars are primarily used in the crossbreeding programs, often through crossing with Landrace or Yorkshire

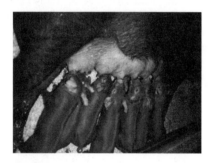

FIGURE 10–10 Duroc piglets

FIGURE 10–11 Landrace sow (Courtesy of American Landrace Association, Inc., IN)

FIGURE 10–12 Yorkshire boar (Courtesy of American Landrace Association, Inc. IN)

sows, both of which have exceptional mothering ability.

Yorkshire (york-shər)   Yorkshires were originally developed in the Yorkshire region of England, and were first imported into the United States in the 1800s. Similar to the Landraces, they are another breed of large white pigs with a long body. Unlike the Landraces, the Yorkshires have erect ears. They grow quickly and efficiently use feed, making them very popular in crossbreeding programs. Qualified animals can be registered with the American Yorkshire Club (see Figure 10-12).

Asian Breeds

Several Asian breeds are receiving attention in the U.S. swine industry in recent years. The breeds are known for having large litters and reaching puberty quickly. They grow more slowly than traditional swine breeds in the U.S. industry, so their use is still limited. For more information on these Asian breeds, and other rare breeds of pigs, refer to the Oklahoma State's Livestock Breeds Web site at www.ansi.okstate.edu/breeds.

WHOLESALE AND RETAIL PORK PRODUCTS

Pork is a very versatile meat, and a wide variety of products are available from swine production. Consumption of pork has remained very stable over the last 10 years, and producers and retailers are seeking ways to increase the market share of pork products (see Table 10-1). Due to aggressive breeding and nutritional adjustments, the pork is even leaner now than it was 10 years ago.

Bacon   The meat product that comes from the sides of the hog. Some of the longer bodied hogs were originally developed to produce maximum amounts of bacon.

Ground pork   Can be from any part of the pig. The meat and trimmings are ground to a uniform consistency.

Hide   The skin of the pig. The hide can be tanned and made into leather for uses similar to the uses of

TABLE 10–1
U.S. per capita food consumption
Meat (individual)
Boneless, trimmed (edible) weight, pounds per capita per year

Year	Pork
2004	47.8
2003	48.4
2002	48.2
2001	46.9
2000	47.8

Figures are calculated on the basis of raw and edible meat. Excludes edible offals, bones, and viscera for red meat. Excludes game consumption for red meat. Calculated from unrounded data.

Source: USDA/Economic Research Service. Last updated Dec. 21, 2005.

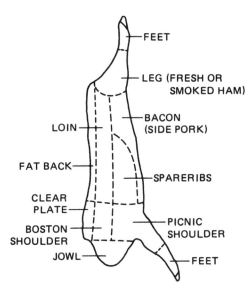

FIGURE 10–13 Primal cuts of the hog carcass

cattle leather. (Footballs are not made of the hide from pigs.)

Lard Hog fat. Historically an income-producing part of the pig, but consumer preference for leaner meat and concerns about consumption of animal fats have greatly decreased the demand for lard.

Primal (prī-məhl) A primal cut is also known as a wholesale cut. Retail cuts are made from primal cuts. Hogs have the following four primal cuts (see Figure 10-14):

  Shoulder primal The shoulder primal includes the front leg and shoulder and is divided into the blade shoulder (top) and picnic shoulder (bottom) **subprimals**. Primary retail cuts from the blade shoulder are the blade (Boston) roast, smoked shoulder roll, blade steaks, and cubes. The primary retail cuts from the picnic shoulder are the picnic roasts, arm steaks, picnic hams, shanks or hocks, and ground pork.

Ham primal The ham primal is the rump and the hind leg of the pig, and is divided in to two subprimals, the butt half and the shank half.

  Butt half Subprimal from the rump. Retail cuts are fresh ham roasts, hams, fresh ham steaks, and center ham slices.

  Shank half The bottom portion of leg. Retail cuts are fresh ham roasts, hams, fresh ham steaks, shanks, or hocks.

Loin The loin is the area from the back of the shoulder to the hind leg along both sides of the backbone. The loin is divided into the following three subprimal cuts:

  Rib The subprimal nearest the shoulder. Primary retail cuts are rib end roasts, blade chops, rib chops, back ribs, and country-style ribs.

  Center cut The middle portion of the loin primal. The premier retail cuts come from this portion of the hog, including center loin roasts, crown roasts, top loin roasts, racks of pork, tenderloins, steaks, center cut chops, loin chops, back ribs, and Canadian-style bacon.

  Sirloin The last third of the loin primal nearest the rump. This subprimal contains more bone that the rib or center cut subprimals. Primary retail cuts are sirloin roasts, tenderloins, steaks, sirloin chops, sirloin cutlets, and button ribs.

Side or belly primal Directly below the loin primal, the belly primal contains more fat that the loin primal. The following are subprimal cuts:

  Side rib The part of the side that contains the ribs. The primary retail cuts are spare ribs, brisket bone, and St. Louis-style ribs.

  Side pork The remainder of the belly primal after removal of the ribs. The primary retail cuts are side pork, bacon, and salt pork. Side pork is the basic cut, salt pork is side pork that is cured, and bacon is side pork that is cured and smoked.

Specialty meats Specialty meats may not be available in all grocery stores, and may need to be special ordered or purchased from a butcher. Specialty meats include liver, kidney, heart, tongue, pigs' feet, pigs' tails, and pork jowls.

Pork chop A general term for any of a variety of retail cuts from the loin primal. A defining term, such as center cut pork chop, identifies where on the primal the chop originated. A butterflied pork chop is one that is cut laterally partway through the chop, and then laid open.

MANAGEMENT TERMS

The following are terms common in the management of the swine industry:

American Livestock Breeds Conservancy (www.albc-usa.org). An organization dedicated to the protection of, and education regarding, rare breeds of livestock. As the swine industry continues to use more crossbred animals, and develop specific strains of crossbreeds for use, some of the breeds of hogs discussed in this chapter may become more difficult to find.

Backfat The thickness of the fat along the back of the hog.

Biosecurity The process by which steps are taken to minimize the introduction of disease to a farm, and minimize transfer of disease among animals on the farm. Biosecurity is of special interest on confinement facilities. Some facilities have "shower in, shower out" policies, where those visiting the facility must shower before going in, and before leaving.

Boar The male breeding pig. Boars reach sexual maturity at approximately seven months of age, and can be incorporated into the breeding herd.

Boar stud A place for keeping exceptional boars, where semen is collected for shipping to other breeding facilities.

Boar taint (bōr tānt) The odor of pork that is harvested from an adult boar. As boars age, boar taint becomes more pronounced, and results in an unpalatable product for the consumer.

Carcass The animal after slaughter and processing has removed the hide and organs. Ideally, carcass weight is approximately 75 percent of the market weight of the animal (see Figure 10-14).

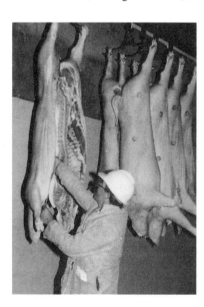

FIGURE 10–14 Carcasses being inspected at a processing plant (Courtesy of Utah Agricultural Experiment Station)

Carcass merit pricing system A pricing system that pays extra to producers for meeting predetermined quality standards. Most hogs are sold on carcass merit pricing systems.

Castration (kahs-trā-shuhn) The removal of the testicles of the male. Castrated male pigs are called barrows.

Checkoff A system where a portion of the sale price of every hog goes to the **National Pork Board** to promote and improve the pork industry. The National Pork Board is responsible for successful promotions such as, "Pork: the Other White Meat," that have helped the pork industry maintain its market share in the face of the increasing consumer demand for poultry.

Closed herd A herd that does not allow new animals on the facility. A closed herd policy reduces exposure of hogs to disease. In a closed herd, an animal is not allowed back on the facility if taken from the facility for any reason.

Confinement system A system in which hogs are raised completely inside buildings (see Figure 10-15).

Contract sales A written contract between a producer and a buyer is established in which hogs are raised to fulfill the contract.

Creep-feeding The practice of providing additional feed to nursing pigs that their mothers cannot reach.

Crossbred seedstock Crossbred animals that are used as breeding animals. The crossbred seedstock is developed from purebred animals.

Cross-fostering Moving piglets from one litter to another to balance litter size. This must be done in the first day or two of life, and with piglets born at the same time.

Direct marketing The most common form of marketing in the U.S. swine industry, in which producers sell hogs to a packing plant or cooperative.

Ear notching A method of putting notches in the ears of piglets to permanently identify them. Most

FIGURE 10–15 Piglets housed in a confinement setting (Courtesy of USDA)

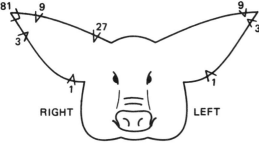

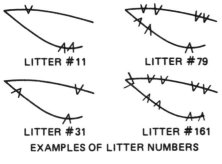

LITTER #11 LITTER #79

LITTER #31 LITTER #161

EXAMPLES OF LITTER NUMBERS

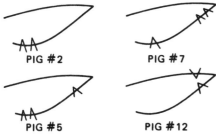

PIG #2 PIG #7

PIG #5 PIG #12

FIGURE 10–16 Ear notching in pigs (Courtesy of Indiana Cooperative Extension Service, Purdue University)

FIGURE 10–17 Farrowing crate (Courtesy of USDA)

producers use a system indicating the litter number, and the individual's number within the litter. Ear notching is the most common method of identifying individual pigs (see Figure 10-16).

Farrowing (fair-row-ing) The delivery of piglets.

Farrowing crate A small pen in which sows are confined to deliver and nurse the piglets. The crate prevents the sows from crushing the piglets by lying on them, but restricts sows from moving freely (see Figure 10-17) and expressing normal behaviors. The use of farrowing crates, also known as **gestation stalls,** has been banned on new facilities in the European Union, and must be completely phased out by 2013. The matter of farrowing crates is a controversial issue in the U.S. swine industry.

Farrowing house A building that is dedicated to the delivery and raising of piglets to weaning.

Farrowing pen An enclosure for females to farrow. As opposed to farrowing crates, farrowing pens allow more expression of natural instincts related to farrowing, such as nesting behaviors.

Feed efficiency The pounds of feed required for each pound of gain.

Feeder pig A pig that weighs 35–50 pounds and is ready to be fed to market weight. Pigs that are part of a **farrow-finish operation** will be fed to slaughter weight on the same facility where they were born. If the feeder pigs were produced at a **feeder pig production facility,** they will be sold and shipped to another facility to be fed to market weight.

Grading A system by which USDA inspectors determine the quality of a carcass based on the amount of backfat and the amount of muscling. Grades range from U.S. No. 1 (the highest grade), to U.S. Utility (the lowest grade) (see Figure 10-18). Grading scales are different for barrows and gilts than for slaughter sows. Boars are not graded.

Growing and finishing phase From the nursery phase, pigs move to the growing and finishing phase, where they are raised to full market size.

Hog Belt Traditionally, the area of the Midwest including Iowa, Illinois, and Indiana, where a large percentage of the hogs raised in the United States were concentrated. Hog production has increased in states such as North Carolina and Colorado that are well outside the traditional Hog Belt.

Litter size The number of piglets born in a litter. Litter size is an important performance criterion for sows.

Loin eye The cross section of the loin muscle that is seen in retail cuts such as the pork chop. The loin eye is the optimal part of the cut, so a larger loin eye is preferred.

Market animal An animal that will be raised to market weight and sent to slaughter. All crossbred barrows and most crossbred gilts are market animals.

Market weight The weight at which hogs are sent to slaughter. Most hogs are slaughtered at approximately 260 pounds. The National Pork Board has set a market weight of 260 pounds at an age of 156 days (barrow) or 164 days (gilt) as a goal for the industry.

U.S. NO. 1

U.S. NO. 2

U.S. NO. 3

U.S. NO. 4

U.S. UTILITY

FIGURE 10–18 Feeder pig grading system (Courtesy of USDA)

Maternal lines   Lines of crossbred animals used for breeding that are focused on maternal traits, such as litter size and mothering ability.

Mating systems   Mating systems are different ways that semen is deposited in the reproductive tract of a female. In pen-mating, boars and females are penned together until mating occurs. In hand-mating, one boar and one female are put together when the female is in estrus until mating occurs. The final mating system used in the swine industry is artificial insemination, in which a female is inseminated with semen from a boar, but has no direct contact with the boar. An increasingly popular method of mating in the swine industry, artificial insemination is done with primarily fresh semen, as frozen semen results in lower litter sizes.

Milk replacer   A substitute for milk used to hand-feed orphans or runts.

Needle teeth   Small sharp teeth in the baby pig. These teeth are clipped shortly after birth to prevent the piglet from injuring the sow or littermates (see Figure 10-19).

Nursery phase   The time between the weaning of piglets until they are feeder pig size (around 50 pounds), and moving them to the growing and finishing phase.

Nutrient management   The method of disposing of the waste products of swine production. Large swine operations are defined as confined animal feeding operations (CAFOs), and must follow EPA guidelines regarding the proper disposal of manure and other waste products.

Optimal growth   The fastest growth that is economically efficient.

Pasture system   A system in which animals are raised at least in part out of doors and on grass (see Figure 10-20).

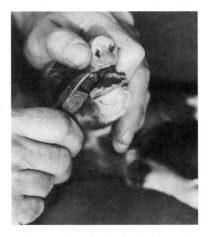

FIGURE 10–19 Clipping the needle teeth of a baby pig (Courtesy of USDA)

FIGURE 10–20 Hogs housed outdoors (Courtesy of Stephen M. Ennis)

Pork Quality Assurance™ Program   A voluntary program for pork producers that certifies producers based on their management practices.

Pork somatotropin (PST) (soh-mah-tah-trō-pihn)   PST is produced by the pituitary and stimulates growth, and increases feed efficiency and carcass leanness. Exogenous PST has not yet been approved in the United States, but has been approved in numerous other countries.

Rate of gain   The number of pounds gained over a fixed period of time, usually on a per day basis.

Replacement gilts   Young females that are retained in the breeding herd.

Runt (ruhnt)   An exceptionally small pig in a litter. Runts often require extra care to thrive.

Shrinkage   The weight loss when a hog is sent to market.

Sire lines   Lines of crossbred animals used for breeding that are focused on growth and carcass traits, such as leanness of carcass, growth rate, and feed efficiency.

Specific pathogen-free hogs   A national system of accrediting that hogs are free of a designated array of parasites and diseases. Accredited facilities must maintain high levels of biosecurity to ensure that the animals are not exposed to pathogens. Regular inspections ensure that all requirements are being met.

Tail docking   The removal of part of the tail of piglets when they are less than three days old. Docking prevents pigs from chewing each other's tails later in life.

Terminal cross   All offspring will go to market and will not be retained for breeding.

Vertical integration   An economic system in which more than one aspect of the production process is owned by the same entity. For example, a packinghouse may also own a finishing facility. Vertical integration can range from one entity owning two segments of an operation, to all segments from seedstock production through processing owned by the same entity.

Weaning (wēn-ing)   The removal of a piglet from the mother, or removal of milk as a nutrient source. In pigs, this may be **early weaning** at 14 days, or traditional weaning age at 21–28 days of age.

Weaning weight   The amount the piglets weigh at weaning.

CHAPTER SUMMARY

The swine industry is a large and economically important part of the animal agriculture industry in the United States. The 1900s was a century of tremendous change for the swine industry, both in structure and product, with a shift from small farms producing fat hogs, to large farms producing large numbers of very lean hogs. Leanness continues to be a priority for the swine industry in response to consumers' desires to have lean meats. The National Pork Board is committed to developing new markets, both at home and abroad, for the pork industry, and checkoff dollars currently assist in that effort. Most market hogs are crossbred animals, but purebred herds are improved to provide quality seedstock for the commercial industry.

STUDY QUESTIONS

Match the breed on the left with its characteristic on the right:

1. _____ Tamworth   a. Black and white with a characteristic belt.

2. _____ Berkshire   b. Developed in Ohio and involved in the development of several other breeds.

3. _____ Duroc   c. Large white hogs known for their mothering ability and floppy ears.

4. _____ Hampshire   d. A Belgian breed with extremely lean carcasses.

5. _____ Landrace   e. Developed in Indiana with contributions from a breed from Gloucester.

6. _____ Poland China   f. The premier bacon breed of Denmark.

7. _____ Yorkshire   g. A red pig with white face and legs.

8. _____ Spotted Swine   h. The pig that had the first organized registry.

9. _____ Hereford   i. The "bacon pig" of Ireland and England.

10. _____ Pietrain   j. A large red pig developed in the eastern United States.

11. A mating that results in a pig that is not going to be used for breeding is called:
    a. Mismating
    b. Market mating
    c. Terminal cross
    d. Dead-end mating

12. What continent is known for breeds of hogs that produce extremely large litters and reach puberty at an early age?
    a. Asia
    b. Africa
    c. Australia
    d. Europe

13. What small sharp teeth are clipped on baby pigs?
    a. Incisors
    b. Fangs
    c. Needle teeth
    d. Molars

14. What is farrowing?
    a. The mating of two pigs
    b. The process of removing part of the tail to prevent chewing
    c. Feeding pigs in a natural setting
    d. The birth/delivery of baby pigs

15. What is the national organization that promotes the swine industry?

16. List three management phases for a hog being prepared for market.

17. What is vertical integration?

Chapter 11
Poultry

Chapter Objectives

▶ Describe how the poultry industry in the United States has evolved in the last 200 years

▶ Understand basic poultry management terminology

▶ Be familiar with the classes and breeds of poultry recognized by the American Poultry Association (APA)

Poultry (pōl' trē) is a broad term used to describe a wide variety of birds that are used in production agriculture. Common types of poultry include chickens, ducks, and turkeys. The American Poultry Association, the first livestock association in the United States, was founded in 1873, to oversee and standardize the exhibition of poultry.

Chickens were first domesticated for cockfighting, a popular spectator sport that has fallen into disfavor over the centuries because it results in the death or maiming of participating birds. Chickens quickly spread across Europe, and many sailing vessels carried chickens to supply eggs and meat for the sailors and passengers. When the Jamestown settlers arrived in the New World in 1607, they brought chickens with them. All poultry were primarily raised on family farms into the twentieth century.

Technological breakthroughs of the 1800s, such as incubators and increased speed of postal delivery, created the poultry industry we see today. Starting in 1918, live chicks could be shipped anywhere in the United States by the U.S. Postal Service. The industry began moving from primarily family farms hatching and raising their own chicks toward a system of a few hatcheries shipping chicks to raisers throughout the country.

The poultry industry is a multibillion-dollar industry in the United States and across the world. In 2005, the poultry industry in the United States was valued at nearly $30 billion (see Figure 11-1). The largest segment of the U.S. poultry industry involves the production of chickens and chicken products. In 2005, the production of eggs was valued at an additional $5 billion dollars (see Figure 11-2). Other avian species that are used in production agriculture include turkeys, ducks, geese, ostriches, and emus.

Birds are also very popular as companion animals. Popular companion birds include parrots, canaries, and parakeets. Chapter 17 discusses these birds more.

CHICKEN BREEDS

All current breeds of domestic chickens are descended from the red jungle fowl (*Gallus gallus*), which still lives in the wild in part of Southeast Asia. The American Poultry Association has divided modern breeds of chickens into different classes, breeds, and varieties. **Class** indicates the region of origin, **breed** indicates a set of birds with similar characteristics, and **variety** indicates a subgroup within a breed, usually based on color or markings. Feather pattern and comb type are two features that may identify different varieties within a breed.

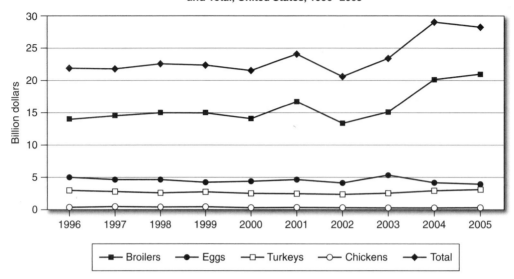

FIGURE 11–1 Total value of poultry production in the United States (Courtesy of USDA)

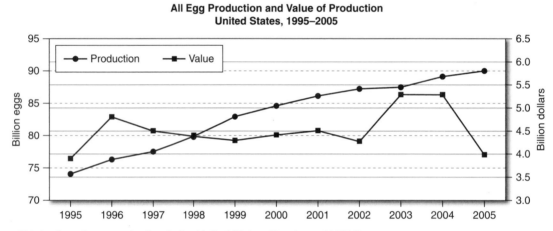

FIGURE 11–2 Total value of egg production in the United States (Courtesy of USDA)

Feather Patterns

The following are feather patterns:

Barred (bahrd)   Feathers with alternating stripes of different colors across the width (see Figure 11-3).

Laced (lāsd)   Feathers with a narrow border of a contrasting color.

Mottled (moth-ld)   When some of the feathers are tipped in white. This differs from spangling in that not all feathers are tipped.

Penciled (pehn-sihld)   Feathers with narrow uniform lines of contrasting color. Depending on the feather and breed, these may be single or multiple.

Spangled (spahng-gehld)   Feathers with a contrasting color at the tip that may be black or white. This differs from mottled in that spangling occurs on every feather.

FIGURE 11–3 The barred feather pattern on a barred rock chicken. (Courtesy of USDA)

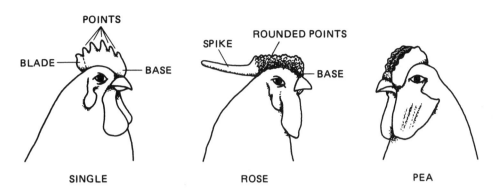

FIGURE 11–4 Types of chicken combs

Comb Types

The **comb** (kōm) is the fleshy growth on the top of the head of chickens. The following are the eight varieties of combs (see Figure 11-4):

Buttercup (buht-tər-kuhp)   A cup-shaped comb centered on the head, with evenly spaced points all around.

Cushion   A comb that is small and close to the head with no spikes, bumps, or other irregularities.

Pea   A comb that is fairly close to the head with three ridges along its length.

Rose   A wide, fleshy comb that is low to the head with a flat top and ends in a spike near the back of the head.

Silkie (sihl-kē)   A round, lumpy comb that may be wider than it is long. This comb type is most often seen on birds with a crest.

Single   A single blade-shaped comb centered on the top of the head, with five or six clearly defined points.

Strawberry   An egg-shaped comb set close to the head.

V-shaped   A comb with two hornlike sections that join in the center of the head and form a definite V shape.

Standard Chicken Breeds

The *Standard of Perfection,* published by the American Poultry Association (APA) since 1874, determines the characteristics for each class and breed of birds in the United States. The APA has determined that "in each breed, the most useful type should be made Standard type." Most standard breeds are of moderate to large size. All of the production birds in the United States are of standard breeds, and most of the bantam breeds have been developed by miniaturizing a standard breed. With the standard size classification, breeds have been divided into classes, based primarily on their place

of origin. The following is a list of the classes and breeds of standard chickens, as defined by the APA *Standard of Perfection.*

American class   Breeds primarily developed in the Americas. The following are American class breeds:

Buckeye   Originally bred in Ohio, the Buckeye is a dark brown dual-purpose bird. Buckeyes were derived from dark Cornish, black-breasted red game, buff Cochin, and barred Plymouth Rock birds.

Chantecler (shahn teh klār)   Originally bred in the province of Quebec in Canada, the Chantecler is a dual-purpose bird that was developed to withstand the Canadian climate. Varieties: white and partridge.

Delaware   Originated in Delaware by crossing barred Plymouth Rocks and New Hampshires, Delawares are dual-purpose birds that grow quickly to broiler size and lay large brown eggs.

Dominique (dohm-ih-nēk)   With unclear origins, a Dominique is a slate-colored, dual-purpose bird that produces brown-shelled eggs.

Holland   A heavy dual-purpose bird that lays white-shelled eggs. White Hollands are derived from white leghorns, Rhode Island Reds, New Hampshires, and Lamonas. Barred Hollands are derived from barred Plymouth Rocks, white leghorns, Australorps, and brown leghorns. Varieties: white and barred.

Java   An ancient breed of poultry that originated in the Far East, the Java was admitted into the *Standard* in 1883. These dual-purpose birds produce eggs with brown shells. Varieties: black and mottled.

Jersey Giant   The Jersey Giant was developed in New Jersey, with an emphasis on producing a large bird with excellent meat-producing qualities. These dual-purpose birds produce eggs with brown shells. Varieties: black and white.

Lamona (lah-mōn-ah)   A dual-purpose bird producing white-shelled eggs, the Lamona was

FIGURE 11–5 New Hampshire chicken (Courtesy of USDA)

FIGURE 11–6 Rhode Island Red (Courtesy of Iowa State University)

developed in Maryland from silver gray dorkings, white Plymouth Rocks, and single-comb white leghorns.

New Hampshire Developed in New Hampshire from original Rhode Island Red stock, the New Hampshire is a dual-purpose bird that produces brown-shelled eggs. Some meat-type strains have been developed within the breed, and many commercial broilers are of New Hampshire bloodlines (see Figure 11-5).

Plymouth Rock One of the first breeds admitted into the *Poultry Standard of Perfection,* the Plymouth Rock is a dual-purpose bird that lays eggs with brown shells, and is used extensively in crossbreeding programs to develop commercial broilers. Varieties: white, buff, silver-penciled, partridge, and Columbian.

Rhode Island Red Originated in Rhode Island from crossing Red Malay game, leghorn, and Asiatic birds, this dual-purpose bird produces brown eggs (see Figure 11-6).

Rhode Island White Similar in type to the Rhode Island Red, Rhode Island Whites were developed by crossing partridge Cochins, white wyandottes, and rosecomb white leghorns. This dual-purpose bird produces brown eggs.

Wyandotte (wy ahn daht) Originated in New York State with the silver-laced variety, the Wyandotte is a dual-purpose bird that produces eggs with brown shells. Varieties: silver-laced, golden-laced, white, buff, black, blue, partridge, and Columbian.

Asiatic class Breeds developed in Asia. The following are Asiatic class breeds:

Brahma (brah-mah) With the original stock coming to the United States from China, and bred with an emphasis on heavy meat production, the Brahma produces brown-shelled eggs and has feathered legs. Varieties: light, dark, and buff.

Cochin (kō-chin) Cochins are of Chinese origin, and are known for their volume of soft plumage

that creates an illusion of great size. Like the other birds in the Asiatic class, Cochins have feathering down the legs and produce brown-shelled eggs. Although originally a meat-type breed, Cochins are now bred primarily for exhibition. Varieties: buff, partridge, white, black, silver-laced, golden-laced, blue, brown, and barred.

Langshan (lang-shahn) Originally bred in China, this dual-purpose bird was developed for both meat and egg production. Langshans have feathering down the legs, and lay dark brown eggs. Varieties: black, white, and blue.

English class Breeds developed in Great Britain and the British Empire. The following are English class breeds:

Australorp (aw-strah lorp) Developed in Australia from the Orpington, the Australorp was bred with the focus on egg production instead of meat production, and was developed as a dual-purpose breed. Australorps produce lightly tinted eggs.

Cornish (korn-ish) Originally developed in Cornwall, England, this large bird is primarily a meat-producing bird, and is used extensively in production of crossbred market poultry. These birds lay brown-shelled eggs. Varieties: dark, white, white-laced red, and buff.

Dorking (dōr king) One of the oldest breeds of domesticated poultry, Dorkings were introduced to England by the Romans. This dual-purpose bird produces white-shelled eggs. Varieties: silver-gray, colored, red, and white.

Orpington (ohr-ping-tehn) Originally bred in Kent, England, the Orpington is a heavy meat-producing bird. Varieties: black, white, buff, and blue.

Redcap Originating in Derbyshire, England, the Redcap is characterized by a very large rosecomb.

They are known as good producers of white-shelled eggs.

Sussex (suhs-ihcks) Originating in Sussex, England, this breed produces dual-purpose birds with an emphasis on meat production. They produce eggs with brown shells. Varieties: speckled, red, and light.

Mediterranean class Breeds developed in countries around the Mediterranean Sea. The following are Mediterranean class breeds:

Ancona (ang-kōnah) An Italian breed similar in type to the Leghorn, Anconas are primarily egg producers, and lay white-shelled eggs.

Andalusian (ahn-dah-loozh-uhn) Developed in the Spanish province of Andalusia, this medium-sized bird produces white-shelled eggs. Variety: blue.

Catalana (kahd-ah-lahn-ah) Developed in Spain near Barcelona, the Catalana is a large dual-purpose bird that produces eggs with white to lightly tinted brown shells. Variety: buff.

Leghorn (lehg-ehrn) Originally from Italy, the leghorn is renowned for its egg-laying capacity, and most commercial egg-laying birds are leghorns. Many varieties of the leghorn were developed in England, the United States, and Denmark. Varieties: This breed has 16 varieties by color and comb type.

Minorca (meh-nhor-kah) The largest of the Mediterranean class of chickens, the Minorca is of Spanish origin and primarily produces white-shelled eggs. Varieties: black (single-comb and rosecomb), white (single-comb and rosecomb), and buff (single-comb).

Sicilian Buttercup (seh-sihl-yan) Developed in Sicily, the Buttercup is most noted for its unique, cup-shaped comb. Buttercups are primarily egg producers, and lay eggs with white shells.

Spanish Developed in Spain, this breed is one of the oldest in the Mediterranean class. The white coloration of the face is a unique characteristic of the breed. Variety: white-faced-black.

Continental class Breeds developed on the European continent. The following are continental class breeds:

Barnevelder (bar-neh-vehl-der) Originally developed in Holland for their brown-shelled eggs, these birds are reddish-brown in color with black accents.

Campine (kahm-pēn) Originally developed in Belgium, the modern Campine is a blend of the two Belgian varieties. These birds are best known as producers of white-shelled eggs, and are popular exhibition birds. Varieties: silver and golden.

Crevecoeur (krev-ker) This crested French breed is very similar in type to the Houdan and the Polish. This dual-purpose bird produces white-shelled eggs. Varieties: black.

Faverolle (faveh-rōl) This French breed was developed primarily as a meat producer. The Faverolle is the only bird in the French breeds that lays brown eggs. Varieties: salmon and white.

Hamburg (hahm-berg) This breed is Dutch in origin, despite its German name. The breed was considerably modified to its current form by breeders in England, where it was popular as a good producer of white-shelled eggs. Hamburgs are now primarily bred for exhibition. Varieties: golden-spangled, silver-spangled, golden-penciled, silver-penciled, white, and black.

Houdan (hoo-dahn) Native to the Normandy region of France, the Houdan is a dual-purpose bird raised for both meat and eggs. Like the Polish and Crevecouer, Houdans have a skull shape that gives the crested appearance. Varieties: mottled and white.

La Fleche (lah-flesh) A French breed with very high-quality meat, this black dual-purpose bird is also a good producer of white eggs. Varieties: black.

Lakenvelder (lah-kehn-vehl-dher) Developed in Germany, these striking birds have black heads and tails, and white bodies. These birds produce eggs that are white to lightly tinted in color.

Polish (pō-lish) Developed in Eastern Europe, this bird is primarily an ornamental and exhibition bird. A unique feature of the Polish, Crevecouer, and Houdan breeds is the shape of the skull, which creates the crested appearance. In addition, some varieties of Polish birds have feathers around the face that may resemble a beard. Polish birds lay white eggs. Varieties: The Polish breed has 11 varieties based on the presence of the beard and color.

Welsummer (well-summer) Developed in Holland from partridge Cochin, partridge wyandotte, partridge leghorn, Barnevelder, and Rhode Island Red stock, this reddish-brown bird is best known for producing brown eggs with spots.

All other standard breeds Breeds that are recognized by the APA but do not fit in any of the other categories. The following are such breeds:

Ameraucana (ah-mehr-ah-cah-nah) Derived from the Araucana, this bird maintains the characteristic blue eggs, and is a dual-purpose bird. Varieties: black, blue, blue wheaten, brown red, buff, silver, wheaten, and white.

Araucana (air-ah-cah-nah) A South American bird also known as the "Easter Egg Chicken" because it lays blue-shelled eggs, this bird is a good dual-purpose chicken. Varieties: black, black red, golden duckwing, silver duckwing, and white.

Aseel (ah-sēhl)   Developed in India, this is an old breed that is known for its aggressive temperament. The birds grow slowly, but produce good quantities of meat and brown-shelled eggs. Varieties: black-breasted red, dark, spangled, white, and wheaten (female).

Cubalaya (kyoo-bah-lāeh)   A Cuban breed descended from Oriental stock, the Cubalaya is prized for its high-quality meat. Varieties: black-breasted red, white, and black.

Frizzle (frihz-ehl)   An exhibition bird of unclear origin, the Frizzle is unique for its curling feathers. They are reasonable producers of brown-shelled eggs. Varieties: clean leg and feather leg.

Malay (mā-lā)   The Malay is originally from Asia and the birds are very large. Malays are long-legged and have contributed to the development of many other breeds in the Standard. The birds are primarily bred for exhibition, and lay dark brown eggs. Varieties: black-breasted red, spangled, black, white, red pyle, and wheaten (female).

Modern game   An exhibition version of the original fighting game chickens of previous centuries, this bird is characterized by its extremely upright carriage and short, tight feathers. Varieties: black-breasted red, brown red, golden duckwing, silver duckwing, birchen, red pyle, white, black, and wheaten.

Naked Neck   Developed in Eastern Hungary and Germany, the Naked Neck has less than half the feathers of a comparably sized bird. They lay brown eggs but were primarily developed as meat producers. Varieties: red, white, buff, and black.

Old English game   The Old English game bird is the descendant of the birds used for cockfighting in past centuries in Britain. The Old English game birds are deeper-bodied than the modern game. Varieties: There are 13 varieties of Old English game birds.

Phoenix (fē-nihks)   A Japanese breed that has been cultivated for nearly one thousand years, the most singular characteristic of the Phoenix breed is the extreme tail growth in the males. Tails have been known to grow in excess of 20 feet. Birds are primarily used for exhibition. Varieties: silver and golden.

Shamo (shā-mō)   A Japanese breed that was used for cockfighting, and now is known as a meat-producing bird. These birds are very tall and lightly feathered, sometimes having exposed skin. Varieties: black, black-breasted red, dark, and wheaten (female).

Sultan (suhl-than)   An ancient ornamental breed originating in Turkey, Sultans have unique feathering on the face, feathers down the legs, and five toes. Females lay white eggs. Variety: white.

Sumatra (seh-mah-trah)   Discovered on the island of Sumatra, this breed has not been crossed with any others, and still appears much like it did at the time of its discovery. Males have long flowing tails, and females produce white or lightly tinted brown eggs. Variety: black.

Yokohama (yō-kah-hahm-ah)   A Japanese breed similar to the Phoenix with the primary characteristic of an extremely long tail. Varieties: white and red shoulder.

Bantams (bahn-tehms)

Bantams are one-fourth to one-fifth the weight of standard birds, with disproportionately large heads, wings, tails, and feathers. In most breeds, weight is the only difference between the bantam and standard types. **Dutch, Japanese, Belgian Bearded d'Anvers** (dē-ahn-vehrz), **Rosecomb, Sebright** (sē' brīt), **Booted, D'uccle** (dē-oohkehl) and **Silkie** (silk' ē) bantams do not have standard breed equivalents. For more information on these unique bantams, refer to the American Poultry Association *Standard of Perfection* (see Table 11-1).

TURKEY BREEDS

Turkeys (ter' kēz)   are the only domestic poultry species of North American origin. Early explorers took wild turkeys back to Europe, where they were domesticated in several European countries by the 1500s. When colonists settled back in North America, they returned the domestic turkey to its native land. The American Poultry Association classifies all turkeys as one breed with various varieties.

Turkey Varieties

The following are varieties of turkeys:

Beltsville small white   A white turkey developed to meet the need for a smaller retail turkey. The male mature weight is 21 pounds, and female mature weight is 12 pounds (see Figure 11-7).

Black   A turkey with solid black throughout. The male mature weight is 33 pounds, and female mature weight is 18 pounds.

Bourbon red   A turkey that is dark chestnut mahogany in color with a white tail. Male mature weight is 33 pounds, and female mature weight is 18 pounds.

Bronze   A turkey that is coppery bronze in color. The male mature weight is 36 pounds, and female mature weight is 20 pounds.

TABLE 11–1

Breeds of Bantam chickens recognized by the American Poultry Association

Class	Breed	Varieties
Game bantams	Modern game	Birchen, black, black-breasted red, blue, blue-breasted red, brown red, golden duckwing, lemon blue, red pyle, silver duckwing, wheaten, white
	Old English game	Birchen, black, black-breasted red, blue, blue-breasted red, blue golden duckwing, blue silver duckwing, blue brassy-back, brassy-back, brown red, Columbian, Crele, cuckoo, ginger red, golden duckwing, lemon blue, quail, red pyle, self blue, silver duckwing, silver blue, spangled, wheaten, white
Single-comb, clean-legged, other than game bantams	Ancona	single-comb, rosecomb
	Andalusian	Blue
	Australorp	Black
	Campine	Golden, silver
	Catalana	Buff
	Delaware	Single-comb
	Dorking	Colored, silver-gray
	Dutch	Light brown, silver, blue, light brown
	Frizzle	All single-comb breeds
	Holland	Barred, white
	Japanese	Black, black-tailed buff, black-tailed white, brown red, gray, mottled, wheaten, white
	Java	Black, mottled
	Jersey Giant	Black, white
	Lakenvelder	Single-comb
	Lamona	White
	Leghorn	Barred, black, black-tailed red, buff, Columbian, dark brown, golden duckwing, light brown, red, silver, white
	Minorca	Black, buff, white
	Naked Neck	Black, buff, red, white
	New Hampshire	Single-comb
	Orpington	Black, blue, buff, white
	Phoenix	Golden, silver
	Plymouth Rock	Barred, black, blue, buff, Columbian, partridge, silver-penciled, white
	Rhode Island Red	Single-comb
	Spanish	White-faced black
	Sussex	Light, red, speckled
	Welsummer	Partridge

(Continued)

TABLE 11–1

Breeds of Bantam chickens recognized by the American Poultry Association *(Continued)*

Class	Breed	Varieties
Rosecomb clean	Ancona	Rosecomb
Legged bantams	Belgian bearded d'Anvers	Black, blue, cuckoo, Mille Fleur, mottled, porcelain, lain, quail, self blue, white
	Dominique	Rosecomb
	Dorking	Rosecomb white
	Hamburg	Black, golden-penciled, golden-spangled, silver-penciled, silver-spangled, white
	Leghorn	Black, buff, dark brown, light brown, silver, white
	Minorca	Black, white
	Redcap	Rosecomb
	Rhode Island Red	Rosecomb
	Rhode Island White	Rosecomb
	Rosecomb	Black, blue, white
	Sebright	Golden, silver
	Wyandotte	Black, blue, buff, buff Columbian, Columbian, golden-laced, partridge, silver-laced, silver-penciled, white
All other combs, clean-legged bantams	Ameraucana	Black, blue, blue wheaten, brown red, buff, silver, wheaten, white
	Araucana	Black, black red, golden duckwing, silver duckwing, white
	Buckeye	Pea comb
	Chantecler	Partridge, white
	Cornish	Black, blue-laced red, buff, dark, mottled, spangled, white, white-laced red
	Crevecoeur	Black
	Cubulaya	Black, black-breasted red, white
	Houdan	Mottled, white
	La Fleche	Black
	Malay	Black-breasted red, black, red pyle, white, spangled, wheaten (female)
	Polish	Bearded buff-laced, bearded golden, bearded silver, bearded white, nonbearded buff-laced, nonbearded golden, nonbearded silver, nonbearded white, nonbearded white crested black, nonbearded white crested blue
	Shamo	Wheaten, black, dark
	Sicilian Buttercup	
	Sumatra	Black, blue
	Yokohama	White, red-shouldered

TABLE 11–1 *(Continued)*

Class	Breed	Varieties
Feather-legged bantams	Booted	Nonbearded black, nonbearded Mille Fleur, nonbearded porcelain, nonbearded, self blue, nonbearded white
	Belgian bearded d'Uccle	Bearded black, bearded golden neck, bearded Mille Fleur, bearded mottled, bearded porcelain, bearded self blue
	Brahma	Buff, dark, light
	Cochin	Barred, birchen, black, blue, brown red, buff, Columbian, golden-laced, mottled, partridge, red, silver-laced, White
	Faverolles	Salmon, white
	Frizzles	Feather-legged
	Langshans	Black, white, blue
	Silkies	Bearded black, bearded white, bearded blue, bearded buff, bearded gray, bearded partridge, nonbearded black, nonbearded white, nonbearded blue, nonbearded buff, nonbearded gray, nonbearded partridge
	Sultans	White

FIGURE 11–7 Beltsville Small White Turkey (Courtesy of USDA)

Narragansett (nār-ah-gahn-seht) A turkey that has black feathers with white and slate gray, creating a grayish appearing bird. The male mature weight is 33 pounds, and female mature weight is 18 pounds.

Royal Palm A turkey with distinctive white feathers tipped in black. The male mature weight is 22 pounds, and female mature weight is 12 pounds.

Slate A turkey that is slate blue color throughout. The male mature weight is 33 pounds, and female mature weight is 18 pounds.

White Holland The most common commercial breed of turkey, also known as broad whites or large whites. The male mature weight is 36 pounds, and female mature weight is 20 pounds.

WATERFOWL

Waterfowl are those birds that spend at least part of their life on the water. Ducks and geese comprise the waterfowl recognized by the American Poultry Association. Heavy, medium, and light classes are determined by mature weight of the breed. Ducks also have a bantam class, similar to the bantam class in chickens.

Duck Breeds

The following are the different duck breeds, broken into weight classes:

Heavyweight

Birds in the heavyweight class range from 10 to 12 pounds at maturity for the drakes, and seven to nine pounds mature weight for the mature duck.

Aylesbury (ālz-behr-rhē) A Large white duck (with a mature weight of 10 pounds for males) originally from England with a flesh-colored bill and white skin. This breed is very popular in England as a meat duck, but is behind the Pekin in popularity in the United States. Variety: white.

Muscovy (muhs-kōhv-ē) Originally from South America, the Muscovy duck has distinctive exposed

FIGURE 11–8 A flock of muscovy ducks (Courtesy of USDA)

FIGURE 11–9 White Pekin drake (Courtesy of Dr. C. Darrel Sheraw, Clarion State College)

knobby skin around the face that looks somewhat like red warts (see Figure 11-8). The Muscovy ducks are the only domestic ducks that are not descended from the mallard. Varieties: white, black, blue, and chocolate.

Pekin (pē kihn)   A large white duck (with a mature weight of 10 pounds for males) with a bright orange beak. These ducks are originally from China, and are the primary ducks raised in the United States for meat. Variety: white (see Figure 11-9).

Rouen (roo-ahn)   A large duck (with a mature weight of 10 pounds for males) developed in France, the Rouen is a brown duck with green accents, similar to the wild mallard.

Medium Weight

Birds in the medium weight class range from 7 to 8 pounds for mature drakes, and from six to seven pounds for mature ducks.

Buff   Also known as the Buff Orpington duck, this bird is a medium-sized duck (with a mature weight

of 8 pounds for males) with a uniform buff color throughout. Variety: buff.

Cayuga (kē-oogah)   A medium-sized duck of American origin, the Cayuga is greenish black with a black bill and black legs. Variety: black.

Crested   A medium-sized duck probably of Aylesbury and Pekin descent, this bird has the unique feature of a crest of feathers on the top of the head. Varieties: white and black.

Swedish   A blue-feathered duck that is a good producer of meat, and that forages well. Their bills should be bluish in color. Variety: blue.

Lightweight

Lightweight ducks weigh between 4 and 5 pounds for both mature drakes and mature ducks.

Campbell   A small duck (with a mature weight of 4.5 pounds for males) with a dark green bill and brown-to khaki-colored feathers (see Figure 11-10). Variety: khaki (kah-kē).

Magpie (māhg-pī)   A small duck (with a mature weight of 5 pounds for males) with predominately white feathers. Depending on the variety, these birds have blue or black feathers on the top of the head, and on the back and tail. Varieties: black and white, and blue and white.

Runner   Developed in India, this duck has a long, narrow, cylindrical body and moves with an upright posture. Varieties: fawn and white, white, penciled, black, buff, chocolate, Cumberland blue, and gray.

Bantam Ducks

Call   Developed in England or Holland as live decoys for duck hunting, the Call duck has a distinctive loud voice, and is very round in style. Mature males weigh less than 2 pounds. Varieties: gray, white, blue, snowy, buff, and pastel.

East Indie   This small duck is black to greenish-black with a black bill and black feet. Variety: black.

FIGURE 11–10 Khaki Campbell drake (Courtesy of USDA)

Mallard (mahl-ahrd)   Developed from the wild mallard, the gray variety shows the classic mallard coloring of gray, brown, and green. Varieties: gray and snowy.

Geese Breeds

Geese are the largest of the domestic waterfowl raised in the United States. Although geese have commercial value for meat, feathers, and down, they are primarily raised in the United States for exhibition purposes.

Heavyweight

African   A large goose with a large knob on the beak where the beak joins the head. The color of the knob depends on the variety. Varieties: brown and white.

Embden (ehm-dehn)   A solid white goose with a broad straight orange bill. Variety: white.

Toulouse (too-loos)   A large goose descended from the French greylag, the Toulouse goose has a characteristic flat bill and a dewlap extending from the bottom jaw to the neck. Varieties: gray and buff.

Medium Weight

American buff   A medium-sized goose with an orange bill and legs, and buff-colored throughout. Variety: buff.

Pilgrim   Unique to the Pilgrim, the color is linked to the gender. Male Pilgrims are creamy white, and females are olive gray (see Figure 11-11). Variety: gender-linked.

Sebastopol (she-vahs-the-pōhl)   A medium-sized white goose with distinctive curling feathers on most of the body. Variety: white.

Lightweight

Canada   This gray and black goose has a mature weight of 12 pounds, a black bill, and black legs. Variety: Eastern (common).

FIGURE 11–11 Pilgrim geese (Courtesy of USDA)

Chinese   A small goose (with mature weight at 10 pounds for males) that is bred primarily as an ornamental. These geese have a short body, long neck, and a distinctive knob where the bill joins the head. Varieties: brown and white.

Egyptian   This small goose is reddish-brown, gray, and black, and although widely bred in Africa, is less common in the United States. Variety: colored (brown).

Tufted Roman   A white goose of small size (with mature weight at 12 pounds for males) first bred over 2,000 years ago with a distinctive tuft of feathers on top of the head. Variety: white.

OTHER TYPES OF BIRDS

The following species of birds are a smaller part of the American poultry industry and fit into niches in the industry. Others are raised primarily as hobby animals. Chapter 13 has more information on some of these birds.

Emu (ē-myū)   A large, fast, flightless bird of Australia, sometimes raised for meat and eggs.

Guinea fowl (gihn-ē foul)   The guinea fowl is a small, game-type bird raised primarily as a hobby. The guinea fowl is known for its distinctive shrill cry.

Ostrich (ohst-rich)   A large (up to 6 feet tall and 300 pounds) flightless bird of Africa, the ostrich was originally domesticated for feather production. Ostriches are now also raised for meat, hide, and eggs. The use of their eggs is primarily decorative.

Peafowl   A strictly ornamental bird, the peafowl is usually blue or green in color, with the most distinctive characteristic being the elaborate "eyed" tail of the male peacock.

Pigeon (pihj-ehn)   A small bird with a round body and a small head, the pigeon is raised primarily for hobbies such as racing and showing. A wide variety of pigeons have been bred for specific qualities.

Quail (kwāl)   A small game bird raised for sale to specialty restaurants or for hunting.

Ratite (ra-tīt)   A class of large flightless birds including emus and ostriches that are raised for their meat and hides.

POULTRY PRODUCTS

A wide range of products are created by the poultry industry. For clarity, they have been divided here by the species that produces each type of product.

Chicken Products

Production chickens are of two general types, egg-type chickens and meat-type chickens. Egg-type chickens are also known as **layers.** Female chickens, or **hens,** lay

250 or more eggs per year. The most common breed of commercial egg-laying chicken is the white leghorn. White leghorns lay eggs with white shells and are what people are most accustomed to seeing. However, some breeds lay brown-shelled eggs. There is no difference in the nutritional value of white-shelled versus brown-shelled eggs. Although all laying hens are of basic leghorn type, the commercial poultry industry is much more focused on using strains of breeds in their breeding programs than different breeds. Egg production occurs in most states in the United States (see Figure 11-12).

Meat-type chickens are very different from egg-type chickens. Although they do lay eggs, as do all chickens, they do not lay as many eggs as egg-type chickens. The emphasis in breeding and raising meat-type chickens is to raise chickens that are well muscled, quickly gain weight, and produce high-quality meat for consumption. The overwhelming majority of meat-type chickens are raised as broilers, and so we will refer to the broiler industry for the remaining discussion on meat-type chickens. Broiler production is concentrated in the southeastern part of the United States (see Figure 11-13). The broiler industry is one of a relatively small number of operations with a large number of birds on each operation. The broiler industry is one of the most vertically integrated in animal agriculture, with one company typically owning all aspects of the broiler production process, from breeding to marketing. The following are terms associated with the poultry industry:

Broiler Meat chicken approximately 8 weeks of age and 5.5 pounds.

Eggs Eggs are very high-quality sources of protein and nutrients, and are a staple in many U.S. households (see Table 11-2). Understanding the parts of an egg (see Figure 11-14), and how eggs are processed is important to safe food handling and determining if an egg is safe to eat. The following are the parts of an egg:

Air cell A pocket of air in the wide end of the egg that gets larger as the egg gets older.

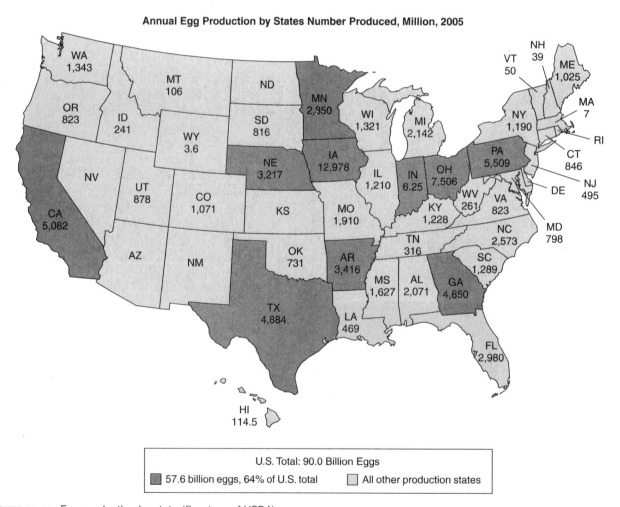

Annual Egg Production by States Number Produced, Million, 2005

U.S. Total: 90.0 Billion Eggs

57.6 billion eggs, 64% of U.S. total   All other production states

FIGURE 11–12 Egg production by state (Courtesy of USDA)

Broiler Production by States Number Raised (000), 2005

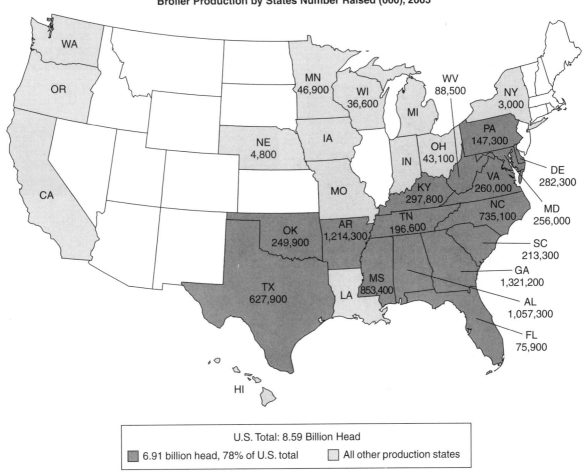

U.S. Total: 8.59 Billion Head

| ■ 6.91 billion head, 78% of U.S. total | □ All other production states |

FIGURE 11–13 Broiler production by state (Courtesy of USDA)

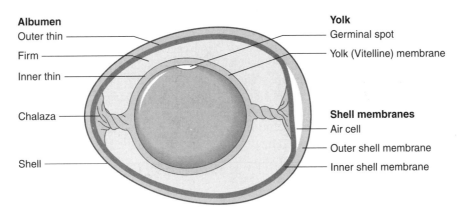

FIGURE 11–14 Parts of the egg

Albumen (al-byū-mehn) The egg white, which is made primarily of water and protein.

Chalaza (keh-lāz-eh) The pair of protein bands that hold the yolk in the center of the egg.

Germ cell The white spot on the yolk that contains genetic material.

Shell Made primarily of calcium, the outer part of the egg that protects the inside of the egg.

Vitelline membrane (vī-tehl-əhn mehm-brān) The thin layer of tissue that holds a yolk's shape.

Yolk (yōk) The yellow part of the egg that contains nutrients.

TABLE 11–2
Egg availability per year

U.S. per capita egg availability			
Farm weight, number per capita per year			
Year	Shell	Processed	Total eggs
2005	175.2	78.8	253.9
2004	179.9	76.2	256.1
2003	182	72.4	254.4
2002	180.1	74.5	254.6
2001	180	72.5	252.5
2000	178	73	251

Calculated from unrounded data.

Source: USDA/Economic Research Service. Last updated Feb. 15, 2007.

Egg grading Eggs are graded to determine their freshness and quality. Many factors are considered when grading eggs, such as the condition of the yolk and albumen. The shell is also examined for flaws and dirt prior to grading the eggs. Eggs are divided into the following three major grades:

Grade AA Grade AA is the highest grade of egg. When viewed from the side, the albumen is firm and creates a raised platform for the yolk, which is also rounded and firm (see Figure 11-15). The Grade AA is the grade of egg that is consistent with being the most freshly laid egg.

Grade A Most eggs in the grocery store are Grade A eggs. Due to the amount of time that passes between the laying of the egg and its appearance in a grocery store, most eggs do not meet the qualifications for Grade AA. Grade A eggs are still very healthy and nutritious, and there should be no concerns about consuming them. In Grade A eggs, the yolk and the albumen are slightly more relaxed (see Figure 11-16), and when viewed from the top, the thick albumen will spread a little farther from the yolk than in Grade AA eggs.

Grade B Grade B eggs are the lowest grade, and are approved only for commercial use and further

FIGURE 11–16 Grade A egg (Courtesy of USDA)

FIGURE 11–17 Grade B egg

processing. They cannot be sold as fresh eggs. In Grade B eggs, the thick albumen has continued to relax, and it may be difficult to differentiate the thick and thin albumen. The yolk also has continued to relax, and is flat rather than rounded when viewed from the side (see Figure 11-17).

Fresh egg An egg produced and sold in its natural form to consumers. Fresh eggs must meet quality standards for commercial sale in the United States.

Roaster Meat chicken approximately 8 weeks of age and greater than 5.5 pounds.

Turkey Products

Turkeys are raised primarily for meat. In 2003, Americans ate 17.4 pounds of turkey per person, according to the National Turkey Federation. Although turkeys used to be eaten primarily at Thanksgiving, vigorous promotion of turkey products in the last 10 years has resulted in a year-round market. Turkeys are still considerably below chicken in the amount of meat consumed per year (see Figure 11-18), but progress is being made by the turkey industry. Turkey operations are more widely spread around the country than the comparable broiler industry (see Figure 11-19).

Waterfowl Products

Ducks and geese are raised primarily for meat for a niche market with a limited economic impact on a national level. There are a few very large duck operations; and the waterfowl industry is still far behind the rest of the poultry industry in regard to consumption and total value of products. However, their feathers and **down** (down)—small, soft fluffy feathers—are used to make pillows, comforters, jackets, and other products. Refer to Chapter 13 for information on the products of alternative poultry species.

FIGURE 11–15 Grade AA egg (Courtesy of USDA)

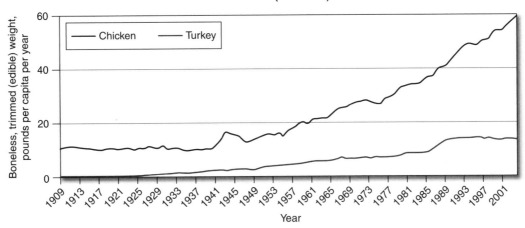

Figures are calculated on the basis of raw and edible meat. Includes skin, neck, and giblets for poultry (chicken and turkey). Excludes use of chicken for commercially-prepared pet food. Calculated from unrounded data.

FIGURE 11–18 Turkey and chicken consumption in the United States (Courtesy of USDA)

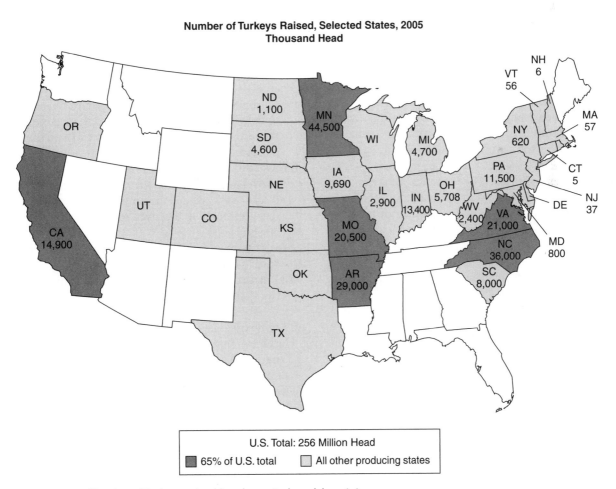

FIGURE 11–19 Number of turkeys raised in primary turky-raising states

MANAGEMENT TERMS FOR POULTRY

All commercial poultry raised in the United States is managed in a similar style; therefore, the following terms will be differentiated by species only if a difference is present.

American Poultry Association   Founded in 1873, the American Poultry Association oversees guidelines for exhibition of poultry, and publishes the *Standard of Perfection,* which describes the characteristics of breeds of poultry in the United States. All descriptions of chickens, turkeys, ducks, and geese in this book come from the *American Poultry Association Standard of Perfection,* 1998.

Average daily gain   The amount of weight an animal gains over time. Calculated by taking a weight at the beginning and end of a period, and dividing the difference by the number of days in the period.

Battery cage   A system of housing poultry in rows of cages in an environmentally controlled building. The size of cages and number of birds per cage is regulated by various federal organizations.

Beak trimming   The practice of removing the tip of the beak on young chicks to prevent chicks from injuring each other (see Figure 11-20).

Bleaching   The loss of pigmentation from the legs and beaks of laying hens that occurs naturally as hens move through the production cycle.

Brooder   A housing unit that includes a heating device to keep young chicks warm.

Brooding/broody/broodiness   Mothering behavior in hens when they are choosing to sit on eggs and incubate them.

Candling (cand-lihng)   The practice of shining a strong light through eggs to evaluate their quality or determine the existence and condition of embryos within the eggs (see Figure 11-21).

FIGURE 11–21 Candling an egg

Cannibalism (kahn-ih-bahl-ihz-uhm)   The practice of one bird attacking another and causing injury. May be one on one, or groups of birds attacking an individual.

Confinement (kohn-fin-mehnt)   The practice of raising birds in a restricted, climate-controlled environment.

Contract growers   People hired by a company to raise a certain number of birds or produce a certain number of eggs for a contracted price.

Coop (koop)   An enclosure to keep birds.

Dressed bird   A bird that has had its feathers removed and is cleaned and ready for food preparation.

Feed efficiency   A calculation of how much feed an animal is consuming to gain a pound of weight. This is calculated by dividing the weight gained over a period of time by the weight of the feed consumed over that period of time.

Flock (flok)   A group of birds.

Force-feeding   The practice of administering food to ducks and geese through a tube to encourage development of the liver for the product foie gras (fwägrä).

Free-range   The practice of raising birds in enclosed yards with shelters (see Figure 11-22).

Hatch (hahch)   The emergence of a baby bird from the shell.

Hatchability (hahch-ah-bihl-ih-tē)   The percentage of fertilized eggs that hatch.

Hatchery (hahch-eh-rē)   A business that sells fertilized eggs ready to hatch, or hatches eggs and sells the chicks.

Hen-day production   Daily rate of egg production reported as a percentage.

Hierarchy (hī-her-ahr-kē)   Also known as "pecking order," hierarchy is a social ranking from most dominant to least dominant that can be linear or nonlinear.

Incubator (ihn-kū-bā-tehr)   The mechanical device that is maintained at the optimal temperature and humidity for the development of embryos and

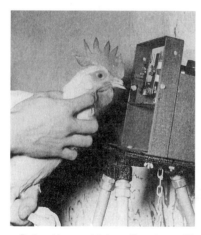

FIGURE 11–20 Debeaking a chicken (Courtesy of Iowa State University)

FIGURE 11–22 Free range chickens

hatching birds. Commercial incubators can hold hundreds of eggs.

Litter   A combination of avian waste and bedding; the bedding before use is known as "fresh litter."

Molt (mōlt)   The natural loss of feathers to allow growth of new feathers. Molting occurs at several stages throughout life. A **forced molt** is the practice of altering the environment to encourage molting, which increases egg production in hens.

Plucking (pluhck-ing)   The process of removing feathers from a bird that may be done manually on a small scale, or mechanically in a large-scale processing plant.

Poults (pōlts)   Baby turkeys until they are sexed and identified as young toms or young hens.

Preening (prēn-ihng)   A behavior in which birds run their beaks across their feathers to distribute beneficial oils throughout the body.

Standard of Perfection   A publication of the American Poultry Association that describes classes and breeds of poultry for exhibition.

Strain (strān)   A related group of animals within a breed. Birds within a strain are more closely related to each other than to other members of the breed.

Tom   An intact male turkey.

Vertical integration   A business practice in which one company manages all levels of production, from hatching through retail distribution.

CHAPTER SUMMARY

The poultry industry is a diverse part of American agriculture with nearly $35 billion generated annually. Participants in the industry range from multinational, vertically integrated corporations that control all aspects of production, to hobbyist breeders and flock owners who raise smaller numbers of birds for personal enjoyment, use, or exhibition. The American Poultry Association recognizes nearly 400 breeds and varieties of poultry, most of which are now used more for exhibition and small flocks than in large commercial flocks.

Consumption of poultry and poultry products has been on a steady increase, and there is no reason to believe that this trend will not continue. Poultry is an affordable source of high-quality protein in the form of both meat and eggs. The poultry industry faces some challenges with animal rights organizations that have concerns about the level of confinement of the birds, and some common management practices. However, the poultry industry is proactively addressing many of these issues as it strives to satisfy consumers in regard to quality of product, as well as management of the birds.

STUDY QUESTIONS

Match the breed of bird with the class to which it belongs. One of the listed classes is used twice.

1. Pekin	a. Mediterranean
2. Khaki Campbell	b. Bantam duck
3. Call	c. Heavyweight goose
4. Leghorn	d. Continental
5. Orpington	e. All other standard
6. Embden	f. American
7. Rhode Island Red	g. Heavyweight duck
8. Old English Game	h. Lightweight duck
9. Lakenvelder	i. English
10. Frizzle	

11. Which of the following is a variety?
    a. Embden
    b. Bantam
    c. Buff
    d. Orpington

12. Which practice is used to reduce damage done through cannibalization?
    a. Candling
    b. Debeaking
    c. Plucking
    d. Molting

13. Which breed constitutes virtually all commercial laying hens?
    a. Rhode Island White
    b. Wyandotte
    c. Ameracauna
    d. Leghorn

14. What organization publishes the *Poultry Standard of Perfection*?

15. How many classes of geese are recognized in the *Standard of Perfection,* and what are they?

16. What English class breed is used extensively in the production of commercial broilers?

17. What is a bantam?

18. What breed of duck is popular for meat production in England?

19. Draw and label the primary parts of the egg.

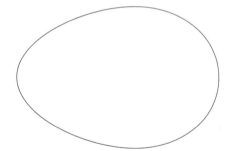

20. The poultry industry is a common target for people opposed to commercial farming. Identify a controversial topic in the poultry industry, research both sides of the controversy, and prepare a short report for your classmates.

Examples of Common Breeds

PLATE 1 Angus Bull. *Courtesy of the American Angus Association.*

PLATE 2 Charolais Bull. *Courtesy of the American-International Charolais Association.*

PLATE 3 American Salers Bull. *Courtesy of the American Salers Association.*

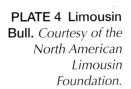

PLATE 4 Limousin Bull. *Courtesy of the North American Limousin Foundation.*

PLATE 5 American Polled Hereford Bull. *Courtesy of the American Polled Hereford Association.*

PLATE 6 American Simmental Bull. *Courtesy of the American Simmental Association.*

PLATE 7 Ayrshire Cow. *Courtesy of the Ayrshire Breeders' Association.*

PLATE 8 Brown Swiss Cow. *Courtesy of the Brown Swiss Cattle Breeders' Association.*

PLATE 9 Guernsey Cow. *Courtesy of the American Guernsey Association.*

PLATE 10 Holstein-Friesian Cow. *Courtesy of the Holstein Association USA, Inc.*

PLATE 11 Jersey Cow. *Courtesy of the American Jersey Cattle Club.*

PLATE 12 Milking Shorthorn Cow. *Courtesy of the American Milking Shorthorn Society.*

PLATE 13 Poland China Swine. *Courtesy of the Poland China Record Association.*

PLATE 14 Duroc Swine. *Courtesy of the United Duroc Swine Registry.*

PLATE 15 Chester White Swine. *Courtesy of Swine Genetics.*

PLATE 16 Landrace Swine. *Courtesy of the American Landrace Association.*

PLATE 17 Hereford Swine. *Courtesy of the National Hereford Hog Record Association.*

PLATE 18 Spotted Swine. *Courtesy of Swine Genetics.*

PLATE 19 Columbia Ram. *Courtesy of the Columbia Sheep Breeders'.*

PLATE 20 Dorset Ram. *Courtesy of the Sheep Breeder Magazine.*

PLATE 21 Hampshire Ram. *Courtesy of the Sheep Breeder Magazine.*

PLATE 22 Rambouillet Ram. *Courtesy of the Sheep Breeder Magazine.*

PLATE 23 Suffolk Ram. *Courtesy of the Sheep Breeder Magazine.*

PLATE 24 Alpine Goat. *Courtesy of the American Dairy Goat Association.*

PLATE 25 Nubian Goat. *Courtesy of the American Dairy Goat Association.*

PLATE 26 Saanen Goat. *Courtesy of the American Dairy Goat Association.*

PLATE 27 Lamancha Goat. *Courtesy of the American Dairy Goat Association.*

PLATE 28 Toggenburg Goat. *Courtesy of the American Dairy Goat Association.*

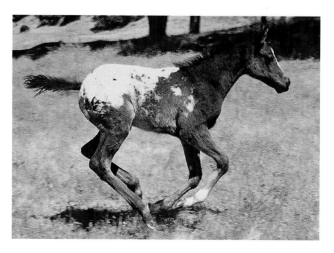

PLATE 29 Appaloosa. *Courtesy of the Appaloosa Horse Club/photo by Tom Poulsen.*

PLATE 30 Quarter Horse. *Courtesy of the American Quarter Horse Association.*

PLATE 31 Arabian. *Courtesy of Johnny Johnston.*

PLATE 32 American Paint. *Courtesy of Don Shugart.*

PLATE 33 Standardbred. *Courtesy of the U.S. Trotting Association.*

PLATE 34 Thoroughbred. *Courtesy of the Illinois Racing News.*

PLATE 35 White Leghorn Rooster. *Photo by Dr. Charles Wabeck.*

PLATE 36 White Plymouth Rock Hen. *Photo by Dr. Charles Wabeck.*

PLATE 37 Broad Breasted Large White Turkey. *Courtesy of the Minnesota Turkey Research and Promotion Council.*

PLATE 38 Broad Breasted Bronze Turkey. *Courtesy of Watt Publishing Co.*

PLATE 39 White Pekin Duck. *Courtesy of Jurgielewicz Duck Farm.*

PLATE 40 Toulouse Goose. *Photo by Dr. Charles Wabeck.*

PLATE 41 English Spot. *Courtesy of the American Rabbit Breeders Association.*

PLATE 42 English Angora. *Courtesy of the American Rabbit Breeders Association.*

PLATE 43 Dutch. *Courtesy of the American Rabbit Breeders Association.*

PLATE 44 Checkered Giant. *Courtesy of the American Rabbit Breeders Association.*

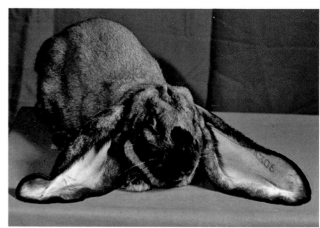

PLATE 45 English Lop. *Courtesy of the American Rabbit Breeders Association.*

Chapter 12
Goats and Sheep

Chapter Objectives

▶ Learn the major sheep and goat breeds

▶ Learn the major products from sheep and goats, and how they are used

▶ Understand the historical role of sheep and goats

▶ Learn the basic management terms relative to sheep and goats

Sheep and goats are a small part of the U.S. animal production industry. Although sheep and goats were among the first animals to be domesticated, and are still a major part of the animal industry in other countries, the consumer demand for product in the United States does not support a large industry (see Figure 12-1). Historically, sheep and goats frequently accompanied explorers and settlers, as their small size and subsequent lower feed needs made them easier to transport than cattle. In the 1800s, the focus of sheep production was on wool; however, as more and more synthetic fabrics have become available, the demand for wool has decreased. Countries such as New Zealand and Australia produce sufficient wool and meat for the needs of American consumers if domestic production ceased.

The sheep and goat industries are more specialty or niche industries, which provide products for specific clientele. Therefore, most producers raise sheep as a hobby or for supplemental income, not as a primary livelihood. One significant part of the industry is the sale of **club lambs,** or lambs intended for exhibition by youth involved in 4-H or Future Farmers of America (FFA) programs. Sheep are an excellent option for youth who wish to keep an animal for a project, but do not have the space or interest in a large animal like a steer or horse. As with swine and cattle, the majority of market lambs are crossbred animals.

SHEEP
Breeds

Although wool is no longer the primary product of sheep in the United States, breeds of sheep are classified based on the type of wool they produce. Although most market animals are crossbred animals, it is important to understand the characteristics of the purebred animals in order to understand what qualities may be found in the crossbred animals.

Fine Wool

Fine wool sheep are bred primarily for wool production, and have lower quality carcasses than meat-type sheep. As wool has become a lower priority product, breeders have improved the carcass quality of fine wool sheep. The following are breeds of fine wool sheep:

Debouillet (deh-boo-lay) A white sheep developed in the United States by breeding Rambouillet and Delaine Merino sheep. The breed does well on the rangelands of the western United States. Debouillets are white-bodied sheep with a white face and legs. They are medium-sized, with adult ewes weighing between 125 and 160 pounds, and adult rams weighing between 175 and 250 pounds.

Merino (meh-rēn-ō) The classic fine wool sheep, the Merino originated in Spain and was imported into

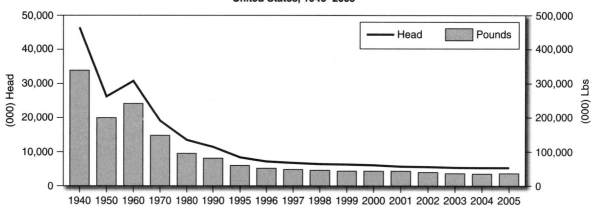

FIGURE 12–1 Comparison of number of sheep and lambs shorn, and pounds of wool produced, in the United States from 1940–2005 (Courtesy of USDA)

the United States in 1801. The Merino has varying degrees of wrinkling to the skin. In the United States, the **Delaine Merino,** the strain with the least wrinkling of the skin, is most preferred (see Figure 12-2). Delaine Merinos are medium-sized white sheep with very uniform fleece quality. Adult ewes weigh between 125 and 180 pounds, and adult rams weigh between 175 and 235 pounds. Size of the sheep is not a primary goal. Merinos are still bred with a focus on producing high-quality wool. Merinos are one of the oldest of sheep breeds, and centuries of selective breeding have resulted in a prepotent sheep that has been frequently crossbred with other breeds to improve their wool characteristics.

Rambouillet (ram-boo-lay) The Rambouillet was developed in France from the Merino and native sheep. Rambouillet sheep are large and have a meatier carcass than Merinos. Mature rams weigh

FIGURE 12–2 A Merino ram (Courtesy of American and Delaine Merino Record Association, OH)

between 250 and 300 pounds, and mature ewes weigh between 150 and 200 pounds. Rambouillet sheep have been used extensively in crossbreeding programs and are the most popular fine wool sheep in the United States.

Medium Wool

Medium wool sheep are primarily raised for meat production. The following are breeds of medium wool sheep:

Cheviot (shehv-ē-əht) A small white sheep developed in northern England's Cheviot Hills, the Cheviot is a hardy sheep that forages well. Documentation of this breed exists from as long ago as 1372. The ewes are known for easy lambing and being good mothers. Cheviots were imported to the United States in the 1830s. The Cheviot has a distinctive white head with white legs and black feet, a stocky body, and moderate fleece quality (see Figure 12-3). The legs and head have no wool covering them. Adult ewes weigh between 120 and 160 pounds, and adult rams weigh between 160 and 300 pounds.

Dorset (dohr-seht) Dorsets are medium-sized white sheep originating in southern England, probably as a mix of native and Merino sheep. Dorsets were originally brought to the United States in 1860, with subsequent importations in the 1880s. Dorsets can be either polled or horned (see Figure 12-4). The polled version was developed in North Carolina, when animals in the purebred herd were born with no horns. Polled Dorsets are the more popular variety in the United States. Dorsets have white legs and faces and produce a good-quality carcass. Adult ewes weigh

FIGURE 12–3 A yearling Cheviot ram (Courtesy of Sheep Breeder Magazine)

FIGURE 12–5 A Finnsheep ewe and her triplets (Courtesy of Bill Carter, Finnsheep Breeders Association, IN)

FIGURE 12–4 A flock of Dorset ewes (Courtesy of Continental Dorset Club, Inc., IA)

FIGURE 12–6 A Hampshire yearling ram

between 150 and 200 pounds, and adult rams weigh between 225 and 275 pounds. Dorsets are one of the breeds that will breed out of season, and can have both spring and fall lambs. The mothers are good producers of milk and often have multiple lambs. The Continental Dorset Club is the registering body for the Dorset sheep.

Finnish Landrace (Fihn-ish lahn-drās) Also known as the Finnsheep, this light-bodied white sheep is from Finland, and was imported into the United States in the late 1960s. Finnish Landraces are renowned for multiple births (see Figure 12-5), and are primarily used in crossbreeding programs to increase the number of lambs born per ewe. Adult rams weigh between 150 and 200 pounds, and adult ewes weigh between 120 and 190 pounds. Animals can be registered with the Finnsheep Breeders' Association, Inc.

Hampshire (hahmp-shər) This medium-bodied sheep originating from southern England has black legs, as well as a black face and ears. They have some wool on the face, but it must not come down the face to the level of the eyes. They may also have some wool on the lower legs, but this is discouraged. They are large sheep with blocky bodies and are naturally polled (see Figure 12-6). Adult rams weigh 275 pounds or more, and the adult ewes weigh 200 pounds or more. Hampshires grow quickly and efficiently convert feed into marketable meat. They were imported into the United States in the 1800s and are very popular in crossbreeding programs to produce market lambs.

Montadale (mon-tah-dāl) The Montadale is a medium to large breed that is all white (see Figure 12-7). Montadales were developed in the 1900s in Missouri by crossing Cheviots and Columbias. One of the distinguishing characteristics of this breed is a lack of wool on the face and legs, a quality inherited from their Cheviot foundation stock. Montadales are naturally polled and produce a good-quality carcass. Adult males weigh between 200 and 275 pounds, and adult females weigh between 150 and 200 pounds. Montadales produce higher-quality wool than many of the other carcass-type breeds, and are known as a

FIGURE 12–7 A Montadale ram lamb

FIGURE 12–8 An Oxford ram (Courtesy of American Oxford Sheep Association, IL)

FIGURE 12–9 A Shropshire ram lamb (Courtesy of Sheep Breeder Magazine)

FIGURE 12–10 A Southdown ram (Courtesy of American Southdown Breeders Association, TX)

dual-purpose breed. Animals can be registered with the Montadale Sheep Breeders' Association.

Oxford   A large sheep that was developed in England in the 1830s from combining Hampshire, Cotswold, and Southdown breeds. The Oxford is a very large sheep, and ewes are prolific producers, averaging 1.5 lambs per year. Adult rams weigh between 200 and 300 pounds, and adult ewes weigh between 150 and 200 pounds. This sheep has a blocky body, with grayish-brown legs, face, ears, and nose. The wool on the Oxford grows down over the poll and between the eyes (see Figure 12-8). They are popular in crossbreeding programs because they produce lambs of good size, and the ewes provide plenty of milk.

Oxfords do best in programs where plenty of feed is readily available, and where they do not need to forage for feed. Animals can be registered with the American Oxford Sheep Association.

Shropshire (shrahp-shər)   The Shropshire was developed in England in the 1860s and was quickly imported into the United States. The hardiness and adaptability of Shropshires made them very popular with the colonists and on small-family farms. They are a relatively small breed, but grow quickly to market weight. Adult rams weigh between 225 and 250 pounds, and adult ewes weigh between 150 and 180 pounds. These sheep are naturally polled, have dark faces and legs, and wool on the face (see Figure 12-9). Because of their relatively small frame, Shropshires that are fed to heavier market weights have a tendency to become fat. The U.S. market prefers a larger animal, and Shropshires are a relatively minor breed in the United States.

Southdown   The Southdown was developed in southern England in the 1700s, and many breeds have at least some Southdown in their ancestry. Southdowns have gray-brown legs and face, mature quickly, and are naturally polled (see Figure 12-10). These sheep have considerable wool growth on the face, including wool around the eyes. Southdowns are one of the

FIGURE 12–11 A Suffolk ram lamb (Courtesy of The Sheep Breeder and Sheepman, MO)

FIGURE 12–12 A Cotswold ram (Courtesy of The Sheep Breeder and Sheepman, MO)

smallest of the medium wool breeds, and grow slowly, so they are not very popular in the United States. Adult rams weigh between 190 and 230 pounds, and adult ewes weigh between 130 and 180 pounds.

Suffolk (suhf-fulk) The Suffolk sheep is the most popular carcass-type sheep in the United States, and is used extensively in crossbreeding programs. The Suffolk was developed in England in the 1800s and was imported to the United States at the end of that century. It is a large, muscular sheep with a black face, legs, and ears. Suffolks have no wool on the legs and are naturally polled (see Figure 12-11). The fast-growing lambs are slow to deposit fat and are very popular market animals. Adult rams weigh between 250 and 350 pounds, and adult ewes weigh between 180 and 250 pounds. Animals can be registered with the National Suffolk Sheep Association.

Tunis (too-nihs) The Tunis was imported into the United States from Africa in the late eighteenth century. The Tunis is a medium-sized sheep with coarse wool and an angular frame. Tunis sheep are unique in that the hair on their clean faces and lower legs is reddish in color. The primary role of the Tunis in the American market is as a producer of lambs for the specialty ethnic market. The Tunis is listed as a rare breed with the American Livestock Breeds Conservancy, but registrations have increased in recent years. Animals can be registered with the National Tunis Sheep Registry, Inc.

Long Wool

Breeds that were originally developed in England, the long wool breeds, have coarse wool and low-quality carcasses. They are usually used in crossbreeding programs. Long wool breeds are less popular in the United States than medium wool breeds.

FIGURE 12–13 A Lincoln sheep (Courtesy of Sheep Breeder Magazine)

Cotswold (kahtz-wohld) The Cotswold is an old English breed that was developed in the 1700s, and imported into the United States in the 1830s. The Cotswold is a polled sheep with long, curling wool that hangs in locks. These locks can be between 8 and 10 inches long. The long wool makes the maintenance of the Cotswold more difficult than some other breeds (see Figure 12-12). The sheep are primarily white, but may have dark spots on the legs. Adult rams weigh around 300 pounds, and adult ewes weigh around 200 pounds.

Lincoln An old English breed, the Lincoln was brought to the United States in the late 1700s. The Lincoln is the largest breed of sheep, with mature rams weighing between 250 and 350 pounds, and mature ewes weighing between 200 and 250 pounds. Lincolns are white with white faces and legs, and are polled (see Figure 12-13). They mature slowly, and have long fleece. The fleece hangs in locks that can be up to 15 inches in length. Although they have large, well-muscled carcasses, Lincolns are best known for their wool production.

FIGURE 12–14 A Romney yearling ram (Courtesy of The Sheep Breeder and Sheepman, MO)

FIGURE 12–15 A pair of Columbia lambs (Courtesy of Columbia Sheep Breeders Association of America, OH)

FIGURE 12–16 A Corriedale ram (Courtesy of Sheep Breeder Magazine)

Romney (rohm-nē) The Romney is a large white sheep from southern England that was imported into the United States in the early 1900s (see Figure 12-14). This is a large breed that does well in difficult conditions. The Romney is primarily a dual-purpose sheep, and lambs grow well on grass to produce high-quality lean carcasses. The Romney has the finest fleece of the long wool breeds, and the fleece is especially popular with craftspeople and those who hand spin wool. Adult rams weigh between 225 and 275 pounds, and adult ewes can weigh between 150 and 200 pounds. Animals can be registered with the American Romney Breeders Association.

Crossbred Wool Breeds

Breeders developed the crossbred wool breeds to create breeds that produce high-quality carcasses and good wool. These breeds were also bred to have a strong flocking instinct, which makes them good breeds for range flocks. The following are crossbred wool breeds:

Columbia The Columbia is a white sheep developed in the United States in the early 1900s (see Figure 12-15). The foundation stock was Lincoln rams and Rambouillet ewes. The Columbia is a large breed and produces relatively lean and good-quality lambs. Columbias are commonly used in crossbreeding programs, especially with some of the black-faced breeds, to produce market lambs. Adult rams weigh between 225 and 300 pounds, and adult ewes weigh between 150 and 225 pounds. Animals can be registered with the Columbia Sheep Breeders Association of America.

Corriedale (cōhr-ē-ah-dāl) The Corriedale is a white sheep that originated in New Zealand in the late 1800s (see Figure 12-16). The foundation stock was Lincoln and Leicester rams, and Merino ewes. The Corriedales produce good-quality fleece, and their carcasses are of moderate quality. Adult rams weigh between 175 and 275 pounds, and adult ewes weigh between 130 and 180 pounds. Animals can be registered with the American Corriedale Association.

Targhee (tahr-gē) The Targhee is a relatively new breed, developed in 1927 in Idaho. The foundation breeds are Rambouillet, Corriedale, and Lincoln stock. Offspring were used in the breeding program based on performance criteria. These white-faced sheep are medium to large and are good sheep for range programs (see Figure 12-17). Mature rams weigh between 200 and 300 pounds, and ewes weigh between 150 and 200 pounds. The U.S. Targhee Sheep Association is the registering organization.

FIGURE 12–17 A flock of Targhee sheep. (Courtesy of U.S. Targhee Sheep Association)

FIGURE 12–18 White Dorper hair sheep (Courtesy of ARS)

Hair Sheep

Hair sheep are a growing segment of the sheep industry. Hair sheep differ from other sheep in that they have more hair follicles, and fewer wool follicles, than wool sheep. As a result, hair sheep do not need to be sheared. The wool price does not justify the expense and effort of shearing for many producers, and raising hair sheep is a viable option. Developed in tropical climates, many hair sheep do well in hot climates and are resistant to many diseases. The following are breeds of hair sheep that may be seen in parts of the United States:

Barbados Blackbelly Barbados Blackbelly sheep are covered uniformly with brown hair, and have black hair on the legs, nose, forehead, and ears. Males have long hair that hangs under the neck and down over the brisket. Rams and ewes are polled. They are lightly muscled, especially in the hindquarter.

Dorper The Dorper was developed in South Africa, and is known as a mutton-type hair sheep. Dorper sheep were developed from the horned Dorset wool breed, and the Blackheaded Persian hair sheep. Dorpers have white bodies, and most have a black head, although the White Dorper has a white head (see Figure 12-18). They are hardy sheep that grow quickly to market weight. The Dorper does well on a range system, and has a highly marketable hide.

Katahdin (kah-tah-dihn) The Katahdin is a hair sheep that was developed in the United States in Maine. Katahdins are hardy sheep that produce meaty carcasses. The ewes lamb easily and are good mothers. Ewes have multiple offspring, and twins can be expected under good management. Katahdins do well in a range or pasture-based management system. Mature rams weigh between 180 and 250 pounds, and mature ewes weigh between 120 and 160 pounds.

St. Croix The St. Croix is also known as the **Virgin Island White** sheep. They were imported into the

FIGURE 12–19 St. Croix hair sheep (Courtesy of ARS)

United States in 1975. St. Croix sheep are adaptable to a wide variety of climates, and forage well (see Figure 12-19). They are very prolific sheep, often having multiple lambs, and having the ability to lamb twice per year. Mature rams weigh 200 pounds, and mature ewes weigh 150 pounds. St. Croix sheep are still rare in the United States.

Other Wool Types

Black-faced Highland An example of a **carpet wool sheep.** The wool of a carpet wool sheep is coarse and long and is best suited for making nonwearable items, hence the name *carpet wool.* The Black-faced Highland is an old breed from Scotland that first came to the United States in the 1860s. These are small sheep with a black face and legs, and long coarse wool. The rams have long curling horns, and the ewes have shorter curved horns. These sheep are not very popular in the United States as a commercial breed, but have some popularity as a novelty breed.

Karakul (kar-ah-kool) Originally from Asia, the Karakul sheep is an ancient breed of **fur sheep** (see Figure 12-20), which means that the pelt of the animal is the primary product. The pelt is harvested from lambs of a few days of age, and is used for coats and other apparel. The wool of the adult is coarse, and the carcass quality is very poor. The Karakul is not popular in the United States.

FIGURE 12–20 Targhee sheep

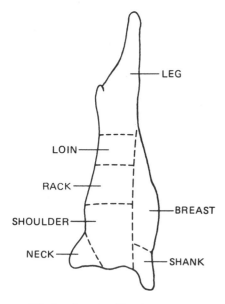

FIGURE 12–21 Wholesale cuts of lamb

Sheep Products

A variety of products are available from the sheep industry. The most consumed in the United States is meat; however, wool, fur, and hides are also marketable products, especially in specialty markets. The following are products from the sheep industry:

Lamb   The meat of a young sheep. Lamb is juicy and flavorful, and popular throughout the world. Just as in other species, the carcass is divided into wholesale cuts (see Figure 12-21).

Breast   The area that includes the lower ribs and abdomen. The breast, rolled breast, and stuffed breast are the retail cuts.

Leg   The leg encompasses most of the hindquarter. Retail cuts include the leg of lamb, the boneless leg of lamb, the sirloin half leg of lamb, and the shank half leg of lamb.

Loin   The loin is the area on the back where there are no ribs attached to the vertebrae. Retail cuts

from the loin include the loin roast, the sirloin roast, the loin chop, and the sirloin chop. The term *lamb chop* refers to any of the chops.

Neck   The neck includes the cervical vertebrae, and does not extend to the scapula. Neck slices are the retail cut.

Shank   The shank is the front leg and includes the elbow. The foreshank is the retail cut.

Shoulder   The shoulder cut includes the scapula, and extends down to the humerus. The retail cuts are the boneless shoulder roast, cushion shoulder roast, square shoulder roast, blade chop, and arm chop.

Rack   The rack includes the ribs. Retail cuts are the rib roast, rib chops, and the crown roast.

Ground lamb   Meat ground to a uniform texture and consistency, analogous to ground beef. Ground lamb can come from any part of the carcass, and can be used in a similar manner to other ground meat.

Milk   Most sheep's milk is made into cheese. Over 50 kinds of cheese worldwide are made from sheep's milk, including **feta** (feh-tah) and **ricotta** (rih-kot-tah).

Mutton (mut-tohn)   The meat of a sheep more than one year of age. Mutton has a reputation for being tough, stringy, and greasy, which has resulted in consumer rejection of the product.

Wool   Economically, wool is a by-product of meat production. Most producers focus on meat production, and income from wool is secondary. Wool is graded based on the quality and texture of the fiber. The following three systems are used in grading (see Table 12-1):

American (blood) system   Wool is graded in a comparative fashion, with the wool of animals of known percentage of Merino blood used as the standards of comparison. This system is losing favor in the industry because it is less precise than other systems of grading.

English (Bradford) system   Wool is graded, and a number assigned based on how much yarn can be spun from one pound of wool. The finer the wool, the more yarn can be spun from each pound.

Micron system   The most objective system of grading, the micron system assigns a grade based on the diameter of the wool fiber in microns.

MANAGEMENT TERMS FOR SHEEP

Accelerated lambing   The management practice of lambing three times in two years (spring, fall, spring). This requires the use of ewes that will breed

TABLE 12–1

Comparison of wool grading systems and the grades of wool produced by different breeds

American blood system	USDA grades	Average fiber diameter (microns)*	Breeds of sheep typically producing grades of wool in range indicated	
Fine	Finer than 80s	17.69 or less	Delaine Merino—	80s or finer
	80s	17.70–19.14	Rambouillet—	64s–70s
	70s	19.15–20.59	Targhee—	62s
	64s	20.60–22.04	Romeldale—	58s–60s
1/2 Blood	62s	22.05–23.49	Corriedale—	60s
	60s	23.50–24.94	Southdown—	56s–60s
3/8 Blood	58s	24.95–26.39	Hampshire, Shropshire—	56s–60s
	56s	26.40–27.84	Suffolk—	54s–58s
1/4 Blood	54s	27.85–29.29	Columbia—	50s–58s
	50s	29.30–30.99	Dorset—	50s–56s
Low 1/4 Blood	48s	31.00–32.69	Cheviot—	48s–50s
	46s	32.70–34.39	Oxford—	46s–50s
Common	44s	34.40–36.19	Tunis, Romney—	44s–48s
Braid	40s	36.20–38.09	Leicester—	40s–46s
	36s	38.10–40.20	Lincoln, Cotswold—	36s–40s
	Coarser than 36s	40.21 or more	Highland—	36s or coarse

*A micron is 1/25, 400 of an inch.

Source: USDA.

out of season, as well as excellent record keeping and management.

Broken mouth The loss of teeth that usually indicates the individual is at least five years of age.

Culling The permanent removal of undesirable animals from a flock.

Dairy sheep Sheep breeds developed with a focus on milk production.

Dipping Immersing an animal in a liquid treatment for external parasites.

Docking Removal of most of the tail shortly after the birth of a lamb (see Figure 12-22).

Drenching A method of administering liquid medication to animals.

Dual-purpose Breeds developed for both meat and wool production.

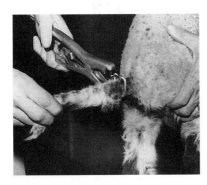

FIGURE 12–22 Docking the tail of a lamb. (Courtesy of USDA)

Early spring lambs Those lambs born in January and February, from ewes that breed early in the breeding season.

Elastrator (ē-lahs-trā-tər) A device used to put a rubber band around the tail for docking, or above the testicles for castration. The rubber band prevents circulation, and causes the tail or testicles to fall off.

Ewe (yoo) Female sheep.

Ewe breeds (yoo) Breeds of sheep with excellent female production characteristics, such as number of lambs per lambing and mothering ability.

Fall lambs Lambs born before December 25, from breeds that will produce "out of season."

Farm flock Sheep raised in a pasture and farm situation. These flocks are raised predominately in the eastern portion of the United States.

Feeder lambs Lambs that are purchased after weaning and are raised and fed to market weight.

Fleece (flēs) The wool removed from a sheep.

Flock expected progeny differences (FEPD) An estimated calculation of the genetic value of every animal in a flock (ram, ewe, or lamb), for which data is submitted to the National Sheep Improvement Program (NSIP). This is valuable for estimating the value of animals within a flock for their contribution to production of the flock.

Flocking instinct The desire of sheep to stay together in a flock is a very valuable quality in sheep that are maintained on rangeland.

FIGURE 12–23 Lamb and ewe in a lambing pen (Courtesy of Iowa State University)

Guard animals   Animals that live with a flock of sheep to protect them from predators. Llamas, donkeys, and specialized breeds of dogs are commonly used.

Hair sheep   Sheep that have fiber that is hair instead of wool. Most hair sheep were developed in warm climates.

Lamb   Baby sheep. For marketing purposes, the following are several classifications of lambs:

   Fed lambs   Lambs that have been raised on grain.

   Hothouse lambs   Lambs less than three months of age.

   Lambs   Any young from seven months to one year old, raised on milk and grass.

   Spring lambs   Lambs between three and seven months old.

Lambing (lahm-ing)   The delivery of lambs by the ewe.

Lambing pen   An enclosure that separates the ewe from the rest of the flock for lambing (see Figure 12-23).

Late spring lambs   Lambs born between March and May. These lambs were conceived at the height of the normal breeding season. Breeding at this time results in the most success getting ewes pregnant, and the most lambs per pregnancy, so late spring lambs are usually the largest percentage of the annual lamb crop. With this large group of lambs coming to market at the same time, market price is often lower for late spring lambs due to the abundant supply.

National Sheep Improvement Program (NSIP)   A federal organization to assist breeders in making mating decisions.

Native ewes   A term used to describe ewes that are best adapted to farm flock conditions. They are generally breeds that produce larger carcasses and more lambs per ewe when compared with **Western ewes.**

Open-face   Sheep that have no wool on the face.

Predators   Animals that kill sheep. Predator control is one of the major challenges for sheep producers.

FIGURE 12–24 Shearing sheep (Courtesy of USDA)

Coyotes are the largest problem in the western states, and dogs are the largest problem in other areas.

Quality grade   As in beef cattle, a quality grade is assigned to sheep. Lambs and yearlings can be graded as (from best to least desirable) **prime, choice, good,** or **utility.** Quality grades for older sheep are **choice, good, utility,** and **cull.** Quality grades are assigned to live animals and to carcasses. Live-quality grade is based on conformation and the amount of finish on the animal, and carcass-quality grade is based on the quality of the meat, the amount of lean meat, and the distribution of fat.

Ram   An intact male sheep.

Ram breeds   Breeds of sheep from which the males are usually used for reproduction. The focus is on meat carcass traits and growth rate of lambs.

Range flock   Sheep raised on the ranges of the western United States.

Shearing (shēr-ing)   The removal of wool from an animal. The wool is removed with a clipper, and should be removed in one piece (see Figure 12-24).

Slaughter sheep   Also called slaughter lambs, these are animals that are being sold directly to slaughter.

Spinning   The practice of transforming wool to yarn.

Staple   The length of the wool fiber. Long staple wool is the most valuable.

Synchronized breeding   Hormonal manipulation of ewes so they all come into estrus at a similar time. This shortens both the breeding season, and the lambing season.

Western ewes   Ewes that are well adapted to range flock conditions. They may have smaller lamb crops, but are hardy and disease-resistant. Western ewes thrive better in farm conditions than native ewes in range conditions.

Wether (weth-ər)   A castrated male sheep (see Figure 12-25).

Wool blindness   A condition in which wool covers the eyes. This condition is of special concern in range sheep.

Wool cap   The wool on some sheep breeds that comes over the poll and down onto the forehead of the sheep.

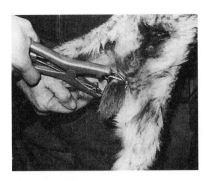

FIGURE 12–25 Following castration, a male lamb is a wether. This lamb is being castrated with an elastrator. (Courtesy of USDA)

Wool cooperative An organization that helps producers sell wool.

Wool pool A method of selling wool by consigning it into a group. Consigned wool is graded and sold based on the grade.

Yearling A sheep between one and two years old. A classification used when marketing sheep.

Yield grade Grading that ranges from yield grade 1 (best) to yield grade 5 (worst), and is based on the amount of backfat.

GOATS

Goats have been domesticated for human use for thousands of years. They produce meat and milk, as well as hide for leather. When explorers were traveling around the world, goats often accompanied them as they were smaller and could produce on a wider variety of forages than cattle. Although the market for goat meat and milk in the United States has historically been low, that is changing. As the country becomes home to more diverse ethnic groups, the market for goat meat is growing.

Goat Breeds

Angora (ahng-gōr-ah) A breed of goat developed in Asia for the production of the hair known as **mohair.** The goats are white, and have no fleece on the face (see Figure 12-26). Most animals are horned. Texas is the leading producer of mohair in the world. Angora goats also produce meat, but the mohair is the primary product.

Boer (bō-ər) A breed of goat from South Africa that was developed for its meat carcass characteristics. The goats have a short white coat with red markings. They are usually horned, although polled individuals are available. The breed was recently imported into the United States, and is growing in popularity both as a purebreed, and for crossing with dairy goats to create a higher-quality carcass.

FIGURE 12–26 Angora goats (Courtesy of Texas A&M University)

FIGURE 12–27 French Alpine doe (Courtesy of Laurelwood Acres, CA)

French Alpine (ahl-pīn) A breed of dairy goat, the French Alpine was imported into the United States in the 1920s. The goats range in color from fawn, red, black, and white, or patterned with white. The ears are erect and they have a short, fine hair coat (see Figure 12-27).

LaMancha (lah-mahn-tcha) The LaMancha is one of the most easily recognized of the goat breeds because of its extremely short ears. This dairy goat can have a **gopher ear,** which is less than one inch in length, or an **elf ear,** which can be up to two inches long. The goat can be any color, and the face has a straight profile. The gopher ear is strongly desired, and bucks with an elf ear are not eligible for registration.

Nubian (new-bē-ahn) The Nubian was developed in Africa. When imported to England, the Nubian was crossed with native dairy goats to produce the breed we see today. The Nubian can be any color, and is characterized by a large drooping ear and a convex **Roman-nosed** profile. Nubians are good producers of milk, and have high butterfat in their milk.

Pygmy goats Pygmy goats are from the Cameroon area of Africa, and were first imported for zoos and other exhibitions. Pygmies are less than 23 inches tall, with relatively short legs and long bodies, and come in a wide variety of colors (see Figure 12-28). They are used for meat and milk, as well as for **pets.**

FIGURE 12–28 Pygmy goat yearling doe (Courtesy of National Pygmy Goat Association)

FIGURE 12–30 Spanish goats (Courtesy of USDA)

FIGURE 12–29 Saanen doe (Courtesy of Laurelwood Acres, CA)

FIGURE 12–31 Toggenburg doe (Courtesy of Laurelwood Acres, CA)

Saanen (sah-nehn)   The Saanen is a cream to white goat from Switzerland that was first imported to the United States in the early 1900s. Saanens have erect ears, most are polled, and both does and bucks may have beards (see Figure 12-29).

Spanish goats   A breed developed as a composite of a variety of dairy and native breeds. Spanish goats are raised primarily for meat production. Unique to some other breeds, Spanish goats will breed year-round, and will have kids twice a year (see Figure 12-30). Spanish goats are being crossed with Boer goats to improve meat production and quality.

Toggenburg (tahg-ehn-berg)   The Toggenburgs are also from Switzerland, and range from fawn to dark brown in color. A distinctive characteristic of this breed is the white line from the eyes down to the muzzle (see Figure 12-31). They may be horned or polled and have erect ears. Some individuals have **wattles,** which is skin hanging below the chin.

Goat Products

Although the market for goat products is small in the United States, some areas of that market are growing. The following are products that are derived from goats:

Butter   Made from the cream of goat's milk. Goat butter is whiter than cow butter.

Cabrito (ka-brē-tō)   The meat of young goats less than 50 pounds.

Cashmere (cazh-mēr)   The fine undercoat of goats. Cashmere can be harvested from any goat. The most desirable cashmere comes from solid-colored animals.

Cheese   The primary product of dairy goat production is milk for cheese. Soft or hard cheeses can be made from goat's milk.

Chevon (shehv-ehn)   Goat meat.

Milk   There is a limited market for fresh or evaporated goat's milk. For some people it is more palatable than cow's milk. Just as in milk from dairy cattle, there is variation in the fat and protein in dairy goat milk based on breed (see Table 12-2).

Mohair (mō-hār)   The hair of an Angora goat. When federal price supports for mohair ended in the 1990s, production decreased.

Management Terms for Goats

Buck   An intact male goat.

Clip   The hair harvested from one animal in one shearing.

TABLE 12–2
Averages of DHI goat herds by Breed. DHI Report K-3, USDA

State	Records		Milk (lb)	Fat (lb)	Protein		(lb)
	Herds	Doe-years			Herds	Doe-years	
Alpine	58	1,769	2,085	69	53	1,684	62
Experimental	4	98	1,528	56	4	98	45
LaMancha	42	714	1,799	66	40	679	55
Mixed Breed	109	6,123	1,760	62	99	5,443	55
Nubian	84	1,121	1,459	66	79	1,084	52
Oberhasli	18	221	1,830	64	18	221	52
Saanen	32	658	1,986	71	32	658	57
Toggenburg	23	301	1,665	57	22	293	47
All Breeds	370	11,004	1,794	64	347	10,159	55

Dairy goats Goats bred primarily for their milk-producing ability.

Doe (dō) A female goat.

Dual-purpose Goats bred for meat and milk production.

Ear notching The practice of making notches in the ear for permanent identification that is most often used with Angora goats raised in range conditions.

Fiber goats Goats bred and kept for their hair.

Kid A baby goat.

Kidding Parturition in goats.

Meat goats Any goats bred for meat production, regardless of breed. Historically, meat goats have not been of a specific breed, but the use of breeds such as the Boer has led to significant improvement in carcass quality.

Milking equipment Large dairy goatherds use mechanical milking equipment similar to those used in dairy cow herds.

Shearing The removal of the fleece from a goat. Angoras are sheared twice annually.

Tattooing Permanently identifying animals with a unique number tattooed in the ear. In LaMancha goats, the tattoo is under the tail.

Wether A castrated male goat.

Management Terms for Sheep and Goats

Browse (browz) The woody twigs and leaves that goats consume as part of their diets. These are generally the younger parts of the plants that are higher in nutritional value.

Flock A group of sheep or goats.

Flushing Increasing the feed for a female prior to breeding to increase reproductive efficiency.

Mixed-grazing The practice of grazing sheep or goats, with cattle, on the same land. Cattle consume the grass, and sheep and goats consume the browse.

Multiple birth The delivery of more than one offspring. Some breeds are more likely to have multiple births than others.

Seasonal breeder An animal that comes into estrus and can be bred only at certain times of the year. Seasonal breeders are often categorized as long-day or short-day breeders, depending on whether they begin to cycle as days grow longer, or as days grow shorter. Sheep and goats are short-day breeders, and are bred in the fall for spring lambs or kids.

Singleton One offspring born.

CHAPTER SUMMARY

The sheep and goat industries are small components of overall animal agriculture in the United States. Although sheep and goats were two of the first species to be domesticated, and are raised throughout the world, consumer demand for these products in the United States is not strong. There is a growing specialty market in areas with ethnic groups that consume goat and sheep products. Meat, milk, and fiber are the primary products of these industries, with mohair being the fiber produced by goats. There is potential for growth in the area of specialty products from these animals.

STUDY QUESTIONS

Match the breed of sheep or goat with the description.

1. _____ Lincoln
2. _____ Angora
3. _____ Nubian
4. _____ Suffolk
5. _____ Katahdin
6. _____ Merino
7. _____ Saanen
8. _____ LaMancha
9. _____ Hampshire
10. _____ Pygmy goat

a. Kept primarily as a pet.
b. The most popular sheep breed in the United States.
c. A hair sheep.
d. An old breed of sheep that was integral in the development of many other breeds.
e. A goat with very small ears.
f. A goat with large, floppy ears.
g. A large sheep with some wool on the face and legs.
h. The largest breed of sheep.
i. A goat raised for its hair.
j. A white or cream-colored dairy goat.

11. Which breed of goat is raised primarily for meat production?
    a. French Alpine
    b. Boer
    c. Toggenburg
    d. Angora

12. What device can be used to castrate lambs?
    a. Elastrator
    b. Browser
    c. Spinner
    d. Tilt table

13. What is the term for a baby goat?
    a. Lamb
    b. Calf
    c. Kid
    d. Wether

14. What are the wool types by which sheep breeds are categorized?

15. What practice involves feeding females more to increase reproductive efficiency?

16. What are the market classifications of lambs?

17. What is a wether?

18. What breed of sheep sets the standard for wool quality?

19. What is culling?

Chapter 13
Alternative Production Animals

Chapter Objectives

▶ Learn about alternative animal agriculture

▶ Learn the species that are primarily involved in alternative animal agriculture

---

Although cattle, swine, and sheep dominate the production animal agriculture industry, a growing number of operations are producing alternative animals for consumption. Marketing is very important for these operations to be successful. For many of the producers, a primary marketing point focuses on the differences in the nutrient profiles of their meat products. The largest of these areas is aquaculture, which is quickly becoming a significant part of the animal agriculture industry.

AQUACULTURE

Aquaculture is the practice of raising fish and related water life for human consumption. A wide variety of aquatic species are raised in aquaculture production, but food fish is the largest component of the industry (see Figure 13-1). The following are some of the aquatic species raised in the aquaculture industry:

Bass   Largemouth bass are farmed primarily as sport fish, for release into recreational fishing areas.

Catfish   The species produced in the greatest quantity in the United States. Catfish have firm white flesh with a mild taste and few bones. There is also a market for catfish for stocking as sport fish.

Mollusks   A class of shellfish including clams, oysters, and mussels that are raised primarily in coastal states.

Salmon   A large fish with reddish flesh. Salmon farming is most prevalent in Washington and Maine.

Shrimp   Shrimp can be farmed in either freshwater or marine (saltwater) environments, depending on the species.

Tilapia (tih-lah-pē-ah)   A white fish that is raised for the commercial fish market.

Trout   A species that has been raised in an aquaculture setting for the longest time. Trout are a medium-sized fish that were originally cultivated for restocking fishing sites, and now are also sold directly for consumption.

Products

Baitfish   Small fish, such as minnows, raised primarily as bait for catching other fish.

Food-size fish   Fish raised to market size for direct sale to consumers.

Stocker fish   Small fish that are sold to operations to be raised to food size, or are reintroduced into a wild environment.

Management Terms for Aquaculture

Blended fishery   An operation that participates in aquaculture and catches fish.

Brackish water   A blend of freshwater and saltwater that usually occurs where freshwater and saltwater bodies meet.

Freshwater   Water with minimal dissolved minerals, such as salt. Rivers and lakes are freshwater.

Marine   Saltwater.

Ocean pens   Enclosures in the ocean for raising fish.

Sport fish   Fish raised and released for recapture by recreational anglers.

Aquaculture Farm Count by Type
U.S. Total—4,309

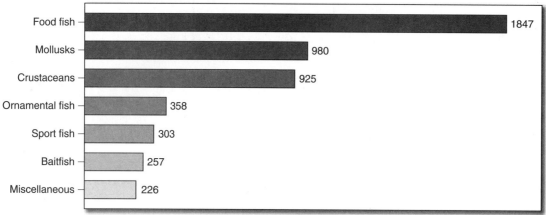

Food fish	1847
Mollusks	980
Crustaceans	925
Ornamental fish	358
Sport fish	303
Baitfish	257
Miscellaneous	226

Source: 2005 Census of Aquaculture, USDA-NASS
Individual category totals do not equal U.S. total since some farms produce multiple poducts.

FIGURE 13–1 Distribution of products of aquaculture farms in the U.S. (Courtesy of USDA)

BISON (BUFFALO)

Bison (bīsuhn) are growing in popularity as sources of nutritious meat. Although the term *buffalo* is frequently used in reference to bison, bison is the most appropriate term. Bison are characterized by a dark brown to black color, a pronounced hump over the shoulders, and a body that tapers toward the hindquarters. Mature males (bulls) weigh up to 2,000 pounds, and mature females (cows) weigh around 1,000 pounds (see Figure 13-2). Bison have one calf per year, in the spring. It is important to remember that bison are basically tamed wild animals; they have not been bred in captivity long enough to share the docile characteristics of domestic livestock. Bison are very territorial, and may aggressively protect that territory, especially during breeding and calving season.

Managing bison has many similarities to managing cattle. Bison are ruminant animals, and can be fed similarly to cattle. Bison are also susceptible to many of the same diseases and parasites as cattle.

Bison have been important sources of food for hundreds of years, ever since Native Americans hunted them for food. After the arrival of European settlers in

FIGURE 13–2 Bison (Courtesy of ARS)

North America, bison were grossly overhunted, nearly to the point of extinction. From a small group of bison that remained, conservationists began to breed bison. The bison that are raised commercially are descended from that group of animals. Bison meat is a red meat that is lower in fat, calories, and cholesterol than beef. Although the bison is a wild animal, the meat does not have a gamy taste like some other wild meat.

Bison Products

Breeding stock Animals sold for breeding purposes. Other animals are sold to hunting facilities or to game preserves or zoos.

Head Sold as a mounted bull's head, or as skulls. Some heads also include all or part of the hump. Heads are primarily used for decorating purposes.

Hide Tanned to make leather and a variety of leather products.

Meat A red meat similar to beef, but lower in fat and cholesterol.

Management Terms for the Bison Industry

Beefalo (beef-ah-lō) A cross between a bison and any breed of cattle.

Breeding age The age at which an animal is recommended to begin breeding. Bison of both genders should be bred no earlier than two years of age. They may reach puberty earlier, but it is advised to wait until the females reach two years of age, and are physically mature, before breeding.

Bulls Male bison.

Calves Baby bison.

Castration Bison are generally not castrated. They reach market weight prior to the onset of sexual behavior, and noncastrated animals grow more quickly than castrated ones. There is no difference in the taste of the meat in castrated and noncastrated males.

Catch pen A small pen that is attached to a large pasture or range that provides a place to confine animals for management techniques. All pens and pastures used to confine bison must be strong and high (six feet recommended).

Cows Female bison.

Dehorning As with other species, bison may be dehorned. Dehorning does reduce incidence of injury to people and other animals, and may result in less damage to hides.

Exotic livestock Nontraditional livestock animals. Regulations regarding slaughter and raising of exotic livestock vary from region to region.

Gestation length The gestation length for bison is 275 days.

Market weight The weight at which an animal is sold for slaughter. For bison, market weight is between 1,000 and 1,200 pounds.

CERVIDAE

Cervidae (ser-vih-dā) is a broad term used to describe many types of deer that may be used as livestock. Regulations controlling the species and use of farmed deer vary, so local and state laws must be considered regarding deer production. Some deer are raised as trophy deer, which means they are raised for hunting. As in the production of deer for meat purposes, it is important to be aware of local and state laws regulating the raising of trophy deer. Varieties used in the United States can be divided into native and nonnative species.

Nonnative Species

Nonnative species are those that are not indigenous to the United States. The following is a list of nonnative species raised in the United States, and the primary products from those species.

Axis deer (ahcks-his) Used for trophy hunts or venison production.

Fallow deer Those primarily raised for venison production.

Red deer A relatively large deer, from which venison and velvet are the primary products.

Sika deer (seek-ah) A relatively large deer, from which venison is the primary product.

Native Species

Elk The largest of the deer, from which velvet, venison, and trophy hunts are the primary products (see Figure 13-3).

FIGURE 13–3 Elk (Courtesy of ARS)

FIGURE 13–4 White-tailed deer (Courtesy of ARS)

White-tailed deer Primarily farmed as trophy animals (see Figure 13-4).

Cervidae Products

Hide The skin of a deer, which can be tanned and processed similar to the hide from cattle.

Ivories The eyeteeth of elk.

Trophy Deer raised for hunting.

Velvet The soft outer covering of the antlers of deer. The velvet is primarily used in the manufacture of medicines for various ethnic groups. Elk can produce up to 16 pounds of velvet per year.

Venison (vehn-ih-zuhn) The meat of deer. Meat from farm-raised deer has a milder taste than wild deer.

Management Terms for Cervidae

Buck A male deer.

Bull A male elk.

Cow A mature female elk.

Doe (dō) A female deer.

Farm-raised An animal that was raised in captivity and fed a controlled diet.

Fawn A baby deer.

Heifer A young female elk.

Single-sire mating The use of one bull elk with a herd of cows.

Multi-sire mating   Multiple bull elk are turned out with a herd of cows, resulting in the exact parentage of offspring possibly being unknown.

LAMAS

Lama (lah-mah) refers to many members of the **camelid** (cam-ah-lihd) family from South America. The **llama** (with two *l*s) and **alpaca** (ahl-pahk-ah) are domesticated, whereas the **guanaco** (gwan-ah-cō) and **vicuna** (vihh-cyun-yah) are still primarily wild animals.

Llamas and alpacas have been part of human history for thousands of years. They were domesticated in Peru about 4,000 years ago. As with many of the domesticated animals in the Old World, llamas and alpacas were multipurpose animals, used for food, fiber, and as pack animals.

Llamas and alpacas were brought to the United States in the late 1800s as unique animals highlighted in zoos and other spectacles. Llamas and alpacas began to find a place in animal agriculture in the United States in the 1970s.

In the United States, they primarily serve as exhibition animals, guard animals (especially llamas) for sheep, and as producers of fiber. There is not a significant market for llama meat in the United States.

Llamas and alpacas are modified ruminant animals, as they have three compartments to the stomach instead of four. Lamas eat forage and will browse, and chew their cud like other ruminant animals. The following are members of the lama family that are domesticated and used in the United States:

Alpaca   The alpaca is the smaller of the domesticated South American camelids, and has shorter, more pointed ears (see Figure 13-5). An adult alpaca weighs between 100 and 175 pounds, and produces about 4 pounds of fiber per year. Alpacas come in a wide range of colors, and are raised primarily for their fleece, for showing, or as pets. The following are alpaca breeds:

Huacaya (hwah-kī-ah)   A breed of alpaca characterized by crimped fleece. The Huacaya is the breed most common in the United States.

FIGURE 13–6 Llamas (Courtesy of USDA)

Suri (suhr-ē)   A breed of alpaca characterized by long, wavy fleece. The fleece can grow almost to the ground.

Llama   The largest of the domesticated South American camelids, the llama has large, banana-shaped ears and comes in a wide variety of colors (see Figure 13-6). Adult llamas weigh between 280 and 450 pounds. Llamas are used for packing and exhibition, and their fleece is sheared and used for cloth. Llamas have coarser fleece than the alpaca. They may also be used as guard animals for sheep.

Camelid Products

Fleece   The hair of a llama or alpaca. The hair is sheared similarly to sheep or Angora goats. The fleece is soft and warm. Alpacas have finer, higher-quality fleece than llamas. Llama and alpaca fleece is also known as **fiber,** and is very popular with people who hand spin fleece for crafts.

Management Terms for Camelids

Berserk Male Syndrome (BMS)   Describes undesirable behaviors that may be seen in male lamas, especially those that are hand-raised. These animals may be excessively aggressive and frequently need to be euthanized because they are unsafe.

Cria (krē-ah)   A baby llama or alpaca (see Figure 13-7).

Fighting teeth   Two sharp canine teeth that erupt in male llamas over two years of age.

Gelding   A castrated male.

Guard llama   An animal that lives with sheep to protect them from predatory animals.

Heat stress   An illness that results from a failure of the body to maintain a normal body temperature.

Herd animal   Llamas and alpacas are herd animals. They must be kept with other llamas and alpacas to flourish.

FIGURE 13–7 Cria (Courtesy of USDA)

FIGURE 13–8 English Angora (Courtesy of American Rabbit Breeders Association)

FIGURE 13–9 Dutch rabbit (Courtesy of American Rabbit Breeders Association)

Induced ovulator An animal that ovulates when mated. Llamas and alpacas are induced ovulators.

Modified ruminant Llamas and alpacas are modified ruminants. Instead of the four compartments in the stomachs of other ruminant animals, llamas and alpacas have three compartments.

Orgle (or-gehl) The call of a male llama or alpaca.

Spitting A behavior in which an animal spits saliva at an animal or person who is perceived to be a threat or an annoyance. Spitting may also occur to establish hierarchy.

Stoic (stō-ihk) A trait of not showing illness or discomfort. Llamas and alpacas are stoic, making it difficult to determine if they are sick.

Stud A male used for breeding.

RABBITS

The rabbit industry is one of the most diverse in animal agriculture. Segments include meat production, fur production, animals for research, and animals for exhibition. Rabbits can be raised on small amounts of property, and are an excellent source of high-quality, low-cholesterol protein. The meat of rabbits is classified as a white meat.

Rabbit Breeds

The American Rabbit Breeders Association (ARBA) is the registering body for rabbits and recognizes more than 40 breeds. For information on all of the breeds, refer to the ARBA Web site (www.arba.net). Breeds are categorized as **six class** or **four class.**

Four-Class Breeds

Four-class breeds are so called because animals for exhibition are divided into four classes: junior buck, junior doe, senior buck, and senior doe. Four-class breeds have an ideal adult weight of less than 9 pounds.

Each breed may have specific ideal weights. Below is a sampling of four-class breeds:

American Fuzzy Lop The American Fuzzy Lop has an ideal adult weight of 3.5 pounds, and may be either solid-colored, or **broken,** a combination of white and another color. Fuzzy Lops have a round head, a long, dense coat that is preferably two inches in length, and ears that hang down by the face. Fuzzy Lops come in a large number of color varieties.

Angora Three breeds of Angora rabbits are recognized by the ARBA: the English, French, and Satin. The Giant Angora is also recognized by the ARBA, but is a six-class breed. All Angoras have long, soft wool that is useful for spinning. The Satin Angora has wool with the translucent qualities of other breeds with **satin** coats (see Figure 13-8).

Dutch A small rabbit, with an ideal exhibition size of approximately 4.5 pounds for adults. The Dutch has short, dense fur, and distinct markings. The rabbit has a white muzzle and blaze, and a white band around the front legs and body (see Figure 13-9).

Florida White The Florida White is the smallest of the meat-type rabbits, with mature adults weighing approximately 6 pounds. Florida White rabbits have stocky, muscular bodies, and uniform white fur. They are also used extensively as laboratory research animals.

Himalayan The Himalayan is a cream-colored rabbit with a darker nose, ears, and feet. Their points can be black, blue, or lilac. The Himalayan is the only rabbit with a cylindrical body type.

Mini Rex A small rabbit, with an ideal adult size of less than 5 pounds, the Mini Rex is one of the most popular show breeds. The Mini Rex has a compact body type, and Rex type fur. There are more than 15 varieties of Mini Rex rabbits.

Netherland dwarf A small rabbit with extremely short ears. The dwarf was bred in the Netherlands by crossing domestic and wild rabbits. With more than 20 color varieties, the Netherland dwarf is a very popular pet and show rabbit.

Tan The Tan rabbit matures to a weight between 4 and 6 pounds. This rabbit has a full-arched body type, and distinctive tan points on the chest, belly, and tail. The tan fur should also rise up from the chest, around the neck, and meet at the base of the ears. Refer to the ARBA standard for more details on the precise markings required in the Tan breed.

Six-Class Breeds

Six-class breeds are larger animals, and have three classifications for each gender: junior, intermediate, and senior. Six-class rabbits have an ideal adult weight of more than 9 pounds; many of the popular breeds raised for meat are members of the six-class group. The following is a sampling of six-class breeds:

Californian The Californian is a white rabbit with a commercial body shape with black feet, ears, tail, and nose. This is one of the most popular meat-type rabbits.

New Zealand The New Zealand is the most popular meat-type rabbit. It has a commercial type body, and can be white, red, or black (see Figure 13-10).

Palomino The Palomino is a moderately large rabbit with a commercial body type. The rabbit can be golden brown, or lynx-colored. Palominos are popular meat-type rabbits.

FIGURE 13–10 New Zealand rabbit (Courtesy of American Rabbit Breeders Association)

Satin The Satin has a commercial body type and is of moderate size. The most distinctive characteristic is the satin fur, which has a unique sheen caused by the presence of translucent guard hairs in the coat.

Rabbit Products

Breeding animals There is a strong market for breeding rabbits of good quality. The market is especially strong for animals of show quality.

Feeder animals Rabbits raised for sale as feed for other carnivores, especially snakes, raptors, or other captive large carnivores.

Meat Rabbit meat is white meat that is low in cholesterol and fat, and high in protein. The following are types of meat rabbits:

   Fryers A young rabbit, usually between 4 and 5 pounds.

   Roaster A mature rabbit.

Research animals Rabbits are often used in a wide variety of research efforts.

Skins Rabbit skins are used in making some clothing.

Wool The hair of some breeds of rabbits. Several breeds produce wool, with the Angora being the most well known.

Management Terms for Rabbits

American Rabbit Breeders Association A national organization that sanctions rabbit shows and trains and certifies judges.

Animal Welfare Act A law outlining requirements for raising rabbits and other animals used in laboratories. Producers raising rabbits for this market must be properly licensed and in compliance with the Animal Welfare Act.

Best of Breed An award given to the best individual in a breed at a show.

Best of Opposite The award given to the best animal that is not the same gender as the winner. For example, if the Best of Breed is a male Mini Rex, the Best of Opposite would be a female Mini Rex.

Best of Show An award given to the best individual of all breeds at the show.

Broken color Any color that has white patches in the pattern.

Buck A male rabbit.

Cecotrophy The consumption of semidigested material that is excreted directly from the cecum. Cecotrophy is different than coprophagy in that the material consumed in cecotrophy is not fully digested fecal material.

Compact body type A short, broad body type that is seen in many of the four-class rabbits.

FIGURE 13–11 Full arched body type in a Checkered Giant (Courtesy of American Rabbit Breeders Association)

Commercial body type   A deep, wide, well-muscled body type. The arch of the back begins just behind the ears. The New Zealand is an example of a commercial body type.

Coprophagy (coh-prof-ah-gē)   The ingestion of feces.

Cylindrical body type   A long, tubular body type that is slightly narrower at the shoulders than at the hindquarters. Himalayans are the only rabbit of this body type.

Dewlap   The fold of skin under the chin. The dewlap is more prominent in some breeds than in others.

Doe   A female rabbit.

Full-arched body type   An arched body type in which the arch begins at the neck and follows through the hindquarter. A Checkered Giant is an example of the full-arched body type (see Figure 13-11).

Holes   Used to describe the number of spots available for rabbits in a carrier or at a rabbitry. A "three-hole carrier" is a rabbit carrier that has separate spaces for three rabbits. A rabbitry that has 50 holes has caging for 50 rabbits.

Hutches   Housing units for rabbits.

Induced ovulator   An animal, such as a rabbit, in which mating stimulates ovulation.

Junior   A young buck or doe.

Kindling (kihnd-lihng)   Parturition in rabbits.

Kindling box   A small box for the female rabbit to nest and deliver her young. Also known as a nest box.

Kits   Baby rabbits.

Malocclusion (mahl-ō-kloo-shuhn)   An often inherited condition, in which the incisors do not meet. Rabbits with malocclusion will have difficulty eating and should not be bred.

Mandolin   Also known as **semi-arch,** mandolin is a body shape that is similar to a pear, with the hindquarters being broader than the forequarter, and the arch beginning at the shoulder.

Nest box   A place for the doe to give birth.

Nesting   A behavior of a doe preparing to give birth, in which the doe prepares the nest box, often even pulling hair from her belly to line the nest box.

Pelleted feed   Commercially available feed that is ground and then recombined in a pelleted feed. The feed can be all concentrate, or a complete feed that contains both the forage and concentrate components of the diet. Pelleted feeds are the most common feed for rabbits. If an all-concentrate pellet is fed, it must be supplemented with hay.

Rabbitry   A facility for raising and housing rabbits.

Rex   A fur type characterized by short, soft, dense fur made from upright hairs. The fur is exceptionally soft, and seen in the Rex and Mini Rex breeds.

Satin   A fur type characterized by a translucent guard hair shaft that gives the coat a unique sheen.

Senior   A mature buck or doe.

Smuts   A crossbreed of Californian and New Zealand rabbits for meat production.

Specialty club   An organization of people with an interest in a single breed. Most rabbit breeds recognized by the ARBA have specialty clubs.

Standard of Perfection   The guide published by the American Rabbit Breeders Association outlining the criteria for each of the accepted breeds and varieties.

Variety   A classification of rabbits within a breed that indicates the color of the animal.

RATITES

Ratites (rah-tīt) are a group of large flightless birds. Ostriches have been raised commercially since the 1800s, when their feathers were the primary product sold. At that time, ostrich feathers were very popular as an accessory for women, who wore them in their hair, or as trim on gowns and other pieces of clothing. The commercial production of ostriches has historically been concentrated in South Africa, but production of all ratites is becoming more widely spread. Three primary species are farmed in the United States: the ostrich, the emu, and the rhea. Cassowaries and kiwis are also members of the ratite family, but are not significantly involved in the U.S. ratite industry. The following are those ratites raised in the United States:

Emu (ē-myoo)   Native to Australia, the emu was first imported for zoos, and has found a specialty market in animal agriculture. Emus reach maturity between 1.5 and 3 years of age. They can either pair mate, or be **polygamous** (poh-lihg-ah-muhs), which means the bird will have several mates. Males are slightly smaller than the females, and

FIGURE 13–12 Emu (Courtesy of USDA)

they have grayish, black, and brown feathers (see Figure 13-12).

Ostrich The largest of the ratites, the ostrich is from Africa and can weigh over 400 pounds. Three varieties (black, blue, and red) are raised commercially. Female ostriches become mature at 2 years of age, and males at 2.5 years of age. Ostriches either pair bond, or mate with several other birds. Although the original reason for raising ostriches was feather production, meat production is now the primary product, and breeding selection is focused on increasing quality and quantity of meat.

Rhea (rē-ah) The rhea is native to South America, and grows to be about 5 feet tall and weighs between 60 and 80 pounds. Rheas are polygamous in their natural environment. Multiple females will lay their eggs in the same nest, and the male will incubate the eggs. Rheas are a pale gray to brown in color, and have no tail feathers.

Ratite Products

Breeding animals Animals purchased, often as pairs, to produce more birds.

Emu oil Produced from the fat of the emu, emu oil is used in pharmaceuticals and cosmetics.

Meat Meat from emus, ostriches, and rheas are red meats, and are lower in fat and cholesterol than beef.

Ostrich feathers Used in feather dusters and for decoration.

Ostrich skin Tanned into leather and used often to make boots and other leather products.

Rhea skin Tanned into leather and used for boots, although a smaller market exists for rhea skin boots than ostrich hide boots.

Management Terms for Ratites

Chick An ostrich from two days to six months of age, or a rhea from time of hatching to six months of age.

Hatchling Ostrich up to two days old.

Hens Female birds.

Yearlings Ostrich from six months to one year of age, and rheas from six months to 18 months of age.

CHAPTER SUMMARY

Although cattle, sheep, and swine are the traditional species used in animal production, a variety of other species have that role. Niche markets exist for products from animals such as deer, bison, rabbits, and ostriches. Much research into the management and sale of products must be done before choosing to raise and market alternative livestock species. Despite the risk of raising alternative livestock, it can be very rewarding for operators with a strong plan. The primary consumer market is people who are interested in the lower-fat meat available from many of the alternative species. However, less infrastructure is needed to support these industries, so operators must be prepared to do much of their own marketing and promotion of the product.

STUDY QUESTIONS

Match the product with the species that produces it. Although more than one species may produce the product, use each product only once.

1. _____ Velvet
2. _____ Venison
3. _____ Feathers
4. _____ Mounted heads with humps
5. _____ White meat

a. Deer
b. Elk
c. Bison
d. Rabbit
e. Ostriches

6. Which is the result of mating a bison with a beef cow?
   a. Cuffalo
   b. Beefalo
   c. Biscow
   d. Cowson

7. Which ratite is not used for production in the United States?
   a. Ostrich
   b. Rhea
   c. Kiwi
   d. Emu

8. What is the form of aquaculture where fish are released into the wild to be recaught?
   a. Food fishing
   b. Mussel fishing
   c. Release fishing
   d. Sport fishing

9. What is the difference between a four-class and six-class breed of rabbit?

10. What products can be sold from ostriches?

11. What is aquaculture?

12. List three species of fish raised for human consumption.

13. Name one market, other than human consumption, for fish raised in aquaculture production.

14. What is the largest deer raised?

15. What is the difference between a llama and an alpaca?

Chapter 14
Horses and Other Equine

Chapter Objectives

► Learn the unique role of horses in human society

► Learn the primary breeds of horses in the United States, and their defining characteristics

► Understand the makeup of the horse industry in the United States

Horses occupy a unique place in human society. In the thousands of years of human life, horses have served every role, from food source to a source of power and transportation, to a partner in recreation. Although the consumption of horse meat is still important in some countries, horses are not consumed in the United States. In some cultures, horses have even been an important part of religious ceremonies. Many horses still occupy traditional work roles, but the majority of horse owners now use their horses for recreational purposes. Like dogs, breeds of horses were developed to do specific tasks, and many breeds were developed within specific regions. It is estimated that over 400 distinct breeds of horses exist worldwide.

Mules and donkeys are also members of the family *Equus,* so are grouped with horses. Horses are of the species *caballus* (kah-bahl-uhs), whereas donkeys are of the species *asinus* (ahs-ī-nuhs). Donkeys are not a breed of horses, but are a different species. Several of the wild asses of Africa are also in the species *asinus.*

BREEDS

Hundreds of breeds of horses are found throughout the world, and many of them can be found in the United States, although in greatly varying numbers. Cross-breeding is less common in the horse industry than in other livestock industries that have been discussed in this text. Grade horses, which are unregistered horses of mixed or pure breeding, do comprise a significant number of animals in the industry, but most breeding programs are focused on the development of purebred animals. Horses are commonly divided into ponies, light breeds, and draft breeds.

Ponies   Ponies are small equines. The upper limit of height depends on breed requirements, but all ponies are less than 58 inches tall at the withers. Some breeds may have height limits below 58 inches. Children often use ponies for driving or for riding. Most ponies are too small for adults to ride. The following are some of the pony breeds that have found popularity in the United States:

Pony of the Americas   The Pony of the Americas (POA) was developed in the United States by crossing a Shetland pony stallion with an Arabian/Appaloosa mare. The ponies are between 46 and 56 inches tall, and have the distinctive Appaloosa markings. These well-muscled, versatile ponies are used for a variety of activities. POA are especially popular with youth, and the POA club has a major focus on serving youth (see Figure 14-1). Ponies are registered with the Pony of the Americas Club.

Shetland   The Shetland is an ancient breed, and one of the smallest ponies. Shetlands get their name from the Shetland Islands off the northern coast of Scotland, where the breed originated. They have been used extensively through history as small draft

FIGURE 14–1 A young girl riding her beloved POA Nugget

FIGURE 14–2 A Welsh Pony (Courtesy of Welsh Pony & Cob Society of America, Inc., VA)

FIGURE 14–3 American Paint Horse Stallion with the tobiano color pattern. (Courtesy of American Paint Horse Association, TX)

horses, especially in the coal mines, where their small size allowed them to fit through the tunnels carved in the earth. Shetlands must stand under 46 inches tall, and may be any color. The American Shetland Pony Club recognizes the following two types of Shetland ponies:

Classic Shetlands A small pony with a short, broad back, round body, and full mane, tail, and forelock. The classic Shetland still looks much like the Shetlands that were first imported to the United States.

Modern Shetland The modern Shetland was developed by crossing classic Shetlands with elegant, high-trotting ponies of other breeds. The resulting pony maintains the small stature of the classic Shetland, with a longer, more upright neck, and more flexion and animation of the knees and hocks than is seen in Classic Shetlands. This "high action" has made Modern Shetlands popular show horses, especially in driving classes.

Welsh (wh-ehlsh) The Welsh breed was present in the mountains of Wales when the Romans invaded Great Britain. These are relatively large ponies, with small-dished faces, long necks, short strong bodies, and excellent legs (see Figure 14-2). The ponies are very athletic, and are used for all activities, from jumping to trail riding and driving. Because they are one of the larger pony breeds, small adults can comfortably ride Welsh ponies. Welsh ponies may be any solid color, but cannot be **piebald** or **skewbald.** The registry, maintained by the Welsh Pony and Cob

Society, is divided into four sections, and variations of the type are registered in each section.

Light horses Light horses are those used primarily for riding or driving carriages and other light wagons. There are hundreds of breeds of light horses throughout the world. Some of the most popular in the United States will be mentioned here.

American Paint Horse The American Paint Horse was developed in the United States by breeding native horses with the distinctive broken color pattern of stock-type horses. The Paint is of stock type, with heavy muscling in the front and rear quarter (see Figure 14-3). The horse should have an attractive head, with a flat or slightly dished profile.

FIGURE 14–4 American Quarter Horse (Courtesy of American Quarter Horse Association, TX)

Although breeders prefer that all animals have the distinctive coat patterns (**overo, tobiano, tovero**), occasionally animals with two registered parents will be born without the pattern. These horses can be registered with the American Paint Horse Association as "**Breeding Stock**" and are not eligible for registration in the regular registry.

American Quarter Horse The American Quarter Horse is the most popular horse in the world, based on total breed registrations. More American Quarter Horses are registered than any other breed. The Quarter Horse originated in the eastern United States from breeding Thoroughbreds to native horses. The horses quickly became famous for their speed over a quarter mile, thus becoming the Quarter Horse. The Quarter Horse moved west with the expansion of the United States, and became the preferred horse for ranch work. The Quarter Horse has an attractive head with a flat, or slightly dished profile and large dark eyes. The Quarter Horse is well muscled throughout the body, especially in the forequarter and hindquarter. Muscling should be round and full (see Figure 14-4). The American Quarter Horse Association, the official registering organization, recognizes 16 acceptable colors; however, the amount of white markings on registered horses is limited.

American Saddlebred The American Saddlebred was developed in Kentucky from Thoroughbred and Narragansett Pacer foundation stock. The goal was to breed a stylish horse that had a smooth and comfortable gait. Morgan and Arabian blood was added, to result in the Saddlebred we are familiar with today. The Saddlebred is the "ultimate showhorse" in the equine world. They stand between 15 and 16 hands tall, with a straight profile, large dark eyes, and a long neck that raises straight up from a well-laid back shoulder. The Saddlebred moves with style and brilliance at all gaits. Saddlebreds can either be three-gaited (walk/trot/canter), or five-gaited (walk/trot/canter/slow gait/rack). Saddlebreds can be almost any color, and are registered with the American Saddlebred Horse Association.

Appaloosa These spotted horses first appeared on ancient Chinese art. However, the Appaloosa breed that we know today was developed on the banks of the Palouse River in Washington State by the Nez Perce Indians, from whom they get their name. The Indians bred the horse to be swift, agile, and useful in war and for hunting. They also bred for the unique spotted pattern. In addition to the spotted coats, Appaloosa characteristics include white sclera around the eye, skin that is mottled pink and black around the muzzle and genitalia, and hooves that are vertically striped dark and light (see Figure 14-5). Appaloosas are similar in type to Quarter Horses, with heavy muscling throughout the body. The Appaloosa Horse Club is the registering organization.

Arabian (uh-rāb-ē-əhn) The Arabian is the oldest of the light horse breeds, and was developed by the **Bedouin** tribes of the Arabian Desert. The Bedouins bred horses to be athletic, beautiful, and to tolerate the harsh desert climate. The Arabian has small sharp ears, a dished face, and large, dark eyes set well on the corners of the head. Arabians are short-backed, with strong bodies, and have characteristic high-tail carriage (see Figure 14-6). According to Bedouin legend, the high-tail carriage was desired to catch the cloak of a warrior if it came off in battle. Arabians

FIGURE 14–5 Appaloosa (Courtesy of Appaloosa Horse Club, Inc., ID)

FIGURE 14–6 Yearling Arabian filly (Courtesy R. Reed)

may be almost any solid color, and are registered in the United States with the Arabian Horse Association. The blood of the Arabian has contributed to the development of almost all light horse breeds in the world, as well as several pony and draft breeds.

Morgan The Morgan horse was developed in the United States in the eighteenth century. All Morgan horses are descended from one horse, Figure, of unknown ancestry. Figure was purchased by a traveling schoolteacher in New England, Justin Morgan, for whom the breed is named. The Morgan is a versatile breed that excels in many equine disciplines. The Morgan is an athletic horse with a straight profile, a long neck, a strong body, and good muscling throughout. Morgans are sound of limb and temperament, and of moderate size, from between 14.1 and 15.2 hands on average. Morgans are most often bay, but can be a wide variety of solid colors. The American Morgan Horse Association is the registering organization.

Mustang (muhs-tang) The Mustang originated in the western United States when horses brought over by the European explorers escaped, and returned to the wild. Because the Mustang is the result of nonselective breeding, color and body type are highly variable.

Standardbred The Standardbred was developed in the United States from breeding Thoroughbreds imported from Great Britain with native horses. Originally, horses were selected based on their ability to **trot** or **pace** the mile in a set (standard) time, hence the name Standardbred. The modern Standardbred is a substantial and athletic horse that excels at racing at a trot or pace in harness (see Figures 14-7 and 14-8). Horses stand between 15 and 16 hands tall, and are most often bay, brown, or black.

Tennessee Walking Horse The Tennessee Walking Horse was developed in central Tennessee in the 1800s by farmers looking for a horse that offered a smooth, comfortable gait suitable to riding all day. The Tennessee Walking Horse was developed by breeding Saddlebreds, Thoroughbreds, and both Canadian and Narragansett Pacers. The resulting horse is of moderate size (15 hands is average), good substance, and of almost any solid color (see Figure 14-9). The Tennessee Walking Horse naturally performs the **running walk,** and even young foals are seen moving through the pasture in this gait. Horses are registered with the Tennessee Walking Horse Breeder's and Exhibitors' Association.

Thoroughbred The Thoroughbred was first created in Great Britain when Arabian horses were bred to native horses. Three Arabian stallions, the Byerly Turk, Godolphin Arabian, and Darley's Arabian are the foundation sires of all modern Thoroughbred horses in the United States and abroad.

FIGURE 14–7 Standard bred trotter "Colonial Charm"

FIGURE 14–8 Standard bred pacer "Besta Bret"

Thoroughbreds are consummate athletes, and excel at all athletic endeavors. They are especially well known for their ability to race and jump. Thoroughbreds are around 16 hands tall, and have long, lean necks, high withers, and well laid-back shoulders. They have long, lean muscling that provides the power for their speed. Thoroughbreds may be any

FIGURE 14–9 Tennessee Walking Horse (Courtesy of Tennessee Walking Horse Breeders and Exhibitors Association, TN)

FIGURE 14–11 Belgian horse (Courtesy of ARS/USDA)

FIGURE 14–10 Thoroughbred horse (Courtesy of ARS/USDA)

solid color (see Figure 14-10). Thoroughbreds have been used extensively to improve other breeds, and many breeds owe at least some of their athleticism to the Thoroughbred. The Jockey Club is the registering organization for Thoroughbreds.

Draft horses Draft horses were developed for pulling heavy loads. They were often crossed with lighter horses to develop heavy carriage horses that could pull a large carriage full of people. With the advent of the tractor, the use of draft horses became much less common in the American countryside. The draft horse is becoming increasingly popular as a show animal, and there is little more dramatic than the sight of a six-horse hitch of draft horses trotting in rhythm. The following breeds are the most common draft horse breeds seen in the United States:

Belgian (behl-juhn) Developed in Belgium, the Belgian is the most popular draft breed in the United States. The horse is well muscled and athletic, and is used in activities ranging from multi-horse hitches to horse pulls, where power is the most important criteria. Belgians are most often golden in color, with a white or blond mane and tail, white markings, and clean legs with no feathering. They are deep-bodied horses with well-sprung ribs and weighing around 2,000 pounds (see Figure 14-11). Belgians are registered with the Belgian Draft Horse Corporation of America.

Clydesdale The Clydesdale is one of the best known of the draft breed, in large part because of the famous "Budweiser Clydesdales" that are exhibited throughout the world. Clydesdales were developed near the River Clyde in Scotland as heavy draft horses. Clydesdales may be more than 18 hands tall, and weigh over 2,000 pounds. They are most often bay or **roan,** with white markings on the legs or face. Clydesdales have moderate **feathering** on their lower legs. Clydesdales are registered with the Clydesdale Breeders of the USA.

Percheron (per-chur-uhn) The Percheron was developed in the Perche region of France, where it got its name. These stylish draft horses stand between 16 and 18 hands tall, and weigh around 2,000 pounds. They have a flat, or slightly dished profile, large dark eyes, large bodies, and well-conformed legs with no feathers. The most common colors are black and grey. Percherons are the most athletic of the draft breeds, and were used extensively as heavy carriage horses in earlier days. They are now also crossed with Thoroughbreds and other riding horses to create athletic horses of size and substance for activities such as **dressage** and jumping. Percherons are registered with the Percheron Horse Association of America.

Shire The shire is the most massive of the draft horse breeds, and the largest breed of horse in the world. Shires were developed in England from the ancient Great Horses that carried knights, and their hundreds of pounds of armor, into battle. Shires stand between 16 and 18 hands tall, and can easily weigh over 2,000 pounds. They may be black, brown, bay, gray, or chestnut with white markings and full feathering on the lower legs. Shires are well muscled throughout the body, and are strong and powerful in their movement. Shires are registered by the American Shire Horse Association.

Suffolk Punch The Suffolk Punch is the least common of the draft horses seen in the United States. This draft breed was developed in the Suffolk region of England. These horses have a rounded, muscular body, and a short, smooth hair coat that is always chestnut in color. The body of these horses is deep and strong through the flank, and the legs are clean of feathers. White markings are minimal, with occasional **stars** and **snips** being the most common. The Suffolk Punch is registered by the American Suffolk Horse Association.

OTHER EQUINE

Donkey The donkey probably originated in Africa, but has been domesticated since before recorded history. The donkey was probably one of the first equine that people used as a work animal. The donkey has a large head, large ears, and a relatively straight neck. The donkey has a tail that is covered with short hair, and with a switch on the end like a cow (see Figure 14-12). Donkeys may range from miniature to mammoth, and are registered accordingly with the American Donkey and Mule Society.

Miniature Horses Miniature Horses were developed from breeding small horses to small horses, with some infusion of Shetland blood. Despite their size (less than 34 inches at the withers), Miniature Horses are not ponies. They should show the same

FIGURE 14–13 Miniature horse pulling a cart

FIGURE 14–14 Meredith Hodges cross-country jumping with her mule. (Courtesy of Meredith Hodges and American Donkey and Mule Society, Lewisville, TX)

balance and structure as full-sized horses, in miniature size. According to the American Miniature Horse Registry, ". . . when looking at a photograph with nothing to give it scale, it should be impossible to tell a Miniature Horse from a full-sized horse" (see Figure 14-13). The Miniature Horse may be any color, or a combination of colors.

Mule The mule is a result of crossing a male donkey and a female horse. The mule has characteristics of both parents, with the longer ears and narrower body of the donkey, combined with the large size, smoother hair coat, and tail of the horse (see Figure 14-14). As with many interspecies crosses, both male and female mules are almost always sterile. A **hinny** (hin-nē) results when a male horse is bred to a female donkey. The characteristics of a mule and hinny are virtually identical, and cannot be differentiated in the mature animals. Mules can range in size from miniature (less than 50 inches tall), to mammoth (more than 68 inches tall). Mules are registered with the American Donkey and Mule Society.

FIGURE 14–12 Two wild jacks registered as standard donkey/wild burros.

EQUINE PRODUCTS

Offspring   The most common product of the equine industry is offspring or breeding animals for sale. Prices range from a few hundred dollars, to millions of dollars, depending on the breed, pedigree, and performance potential of the animal.

Service industry   The service industry is a significant part of the equine industry, and the route for most people making a living in the equine industry. The following are common services offered:

Boarding   Keeping horses owned by others, and providing housing and care for a fee. Board can range from **full board,** which includes all labor and supplies, to **partial board,** where the horse owner provides some labor or supplies. Some horses are on pasture board, which means they live in a pasture, and not in the barn.

Riding instruction   Lessons to teach people how to ride a horse, for which payment is received. This may be instruction with a horse owned by the rider or the instructor.

Training   Care of a horse, including teaching the horse to be ridden, for which payment is received. Training may also include teaching the horse to participate in a particular competitive activity or event. The **trainer** is the person who teaches the horse.

Recreation

Recreation is the primary use for most horses in the United States. The following list is some of the recreational opportunities that are available for participants in the equine industry:

Trail riding   Over 80 percent of people who own horses do so for recreation. One of the most popular recreational activities is trail riding. Trail riding describes casual riding in a noncompetitive environment (see Figure 14-15), such as at state or national parks, or on the home property. Any breed of horse can be used for trail riding, although the **gaited breeds** are often preferred because of their smooth gaits.

Sport

Horses are widely used for competitive activities or sports. Many horse sports originated in activities that used to be part of a horse's work (see Figure 14-16).

Barrel racing   A racing activity in which horses race against a timer to complete a cloverleaf pattern around three barrels (see Figure 14-17). Horses race the course one at a time (see Figure 14-18). The event may be held independently or as part of a rodeo.

FIGURE 14–15 Enjoying a ride on the trail (Courtesy of ARS)

FIGURE 14–16 A group of people preparing to foxhunt. (Courtesy of Library of Congress, Prints and Photographs Division, Theodor Hordczak Collection [reproduction number LC-H824-T-2424-002-xDLC])

Combined driving   An event in which horses are harnessed to a wagon and complete a drive over trails, an obstacle course, and a pattern of cones in an arena.

Dressage (drah-sahj)   A classic style of training that emphasizes the quality of movement of the horse, and teamwork between the horse and rider. Any

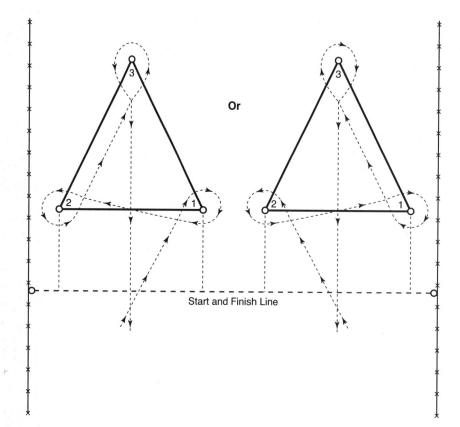

FIGURE 14–17 The barrel racing pattern

FIGURE 14–18 An Appaloosa running the barrel pattern. (Courtesy of Appaloosa Horse Club, Inc., ID)

breed can compete in dressage, and the principles of dressage are valuable for horses ridden in any discipline. Dressage is one of the Olympic equestrian disciplines, and is a component of a three-day event. The United States Dressage Federation (USDF) is the organization that oversees dressage in the United States.

Gymkhana (jihm-kah-nah) Racing events on horseback. Gymkhana includes events such as **barrel racing** and **pole bending** (see Figures 14-17 and 14-19). The horses typically race individually against the clock.

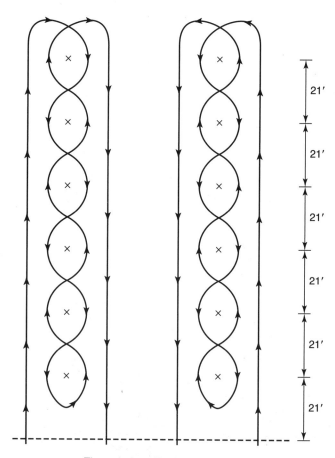

FIGURE 14–19 The pole bending pattern

FIGURE 14–20 Polo

Polo   One of the most ancient horse sports, polo involves teams of riders that compete to knock a ball through goal posts with a mallet (see Figure 14-20). The polo match is broken up into periods called **chukkars** (chuck-ers), similar to innings in baseball, which are 7½ minutes long.

Showing   Any of a wide range of activities in which a horse is exhibited, either ridden, driven, or based on its conformation, and a judge decides which horse most closely fits the standard for that activity. Common show classes are Western Pleasure, English Pleasure, and Showmanship. Each breed has classes that are specific to the breed's origin and unique abilities.

Show jumping   Competition in which horses complete a prescribed course of jumps. Penalties are assessed for knocking down parts of the jumps, or for completing the course in more than the allowed time.

Three-day event   A competition in which horses first complete a dressage phase, followed by a cross-county phase where they must complete jumps over a course of several miles within a prescribed time, and then must complete a show jumping competition. The three-day event is also known as combined training, and is one of the equestrian events at the Olympics.

Racing   Generally refers to an activity in which horses compete directly against one another to complete a distance in the shortest possible time.

    Endurance racing   Races that are up to 100 miles or more in length. Although any breed can participate, breeds such as the Arabian excel.

    Sprint racing   Horses sprint a short distance, ¼ mile or less. Paint Horses, Appaloosas, and Quarter Horses are the primary breeds competing at this distance.

    Thoroughbred racing   Distances range from ½ mile to 2 miles, with most races between ⅞ of a mile and 1¼ miles. The Kentucky Derby, the Preakness Stakes, and the Belmont Stakes (the Triple Crown) are examples of Thoroughbred races.

Reining   A competition in which working ranch horses demonstrate the skills needed by completing maneuvers in a prescribed pattern. Any breed can compete in a reining sponsored by the National Reining Horse Association; however, many breeds have reining as a class at their show, and only horses of that breed may compete in those shows. The following are maneuvers that are combined in different arrangements to make reining patterns:

    Sliding stop   A maneuver in which the horse goes from a gallop to a stop, and slides on its hind feet in a controlled manner.

    Spin   A maneuver in which the hind foot of the horse stays in place, and the horse rotates at a speed to either the left or the right in a complete circle. While spinning, one hind leg stays in place, and the horse crosses its front legs over to spin its entire body around the pivot point created by the stationary hind leg.

Rodeo   A competition in which skills used on the ranch are judged, based on time, points, or a combination of time and points. Horse-related events include:

    Bronc (brohngk) **riding**   An event in which riders ride a horse that tries to buck them off. In **saddle bronc** riding, the horse is equipped with a saddle, whereas in **bareback bronc** riding, there is no saddle. Riders must stay on for eight seconds, and separate points are assigned from a judge for the rider's performance, and the performance of the horse.

    Calf roping   An event in which riders rope a calf from the horse, and restrain the calf as if it was getting a shot or branded. The event is timed, and the rider and horse with the fastest time wins.

    Steer wrestling   An event in which riders gallop next to a steer on their horse, then reach over and dismount their horse to wrestle the steer to the ground. Two horses are involved, one for the rider that wrestles the steer, and a second rider (the **hazer**) on the other side keeps the steer in a straight line. The rider who wrestles the steer to the ground most quickly wins.

Work

Although horses are primarily used for recreation, either competitive or noncompetitive, significant numbers of horses are still used for work in the United States. Some jobs are still better completed by man and

FIGURE 14–21 A child participating in therapeutic riding. (Courtesy of Purdue University)

FIGURE 14–22 A Rangerbred horse showing the blanket spotting pattern. (Courtesy of Colorado Ranger Horse Association, PA)

animal than by man and machine. The following are work-related terms in the equine industry:

Dude ranch A place where people come to enjoy an experience similar to a working ranch, such as riding horses while doing ranch-type work or trail riding.

Police work Many cities use officers on mounted horses for crowd control and other police work.

Ranch work Many ranches still use horses for moving animals, roping and restraining animals for treatment, and separating animals from the herd (**cutting**). Some of the skills for ranch work have evolved into competitive events.

Therapy Horses are being increasingly used as part of physical, mental, and emotional therapy programs. The therapy occurs both through riding (see Figure 14-21) and through programs where the people handle horses. The North American Riding for the Handicapped Association (NARHA) is the major overseeing organization for this type of therapy in the United States. Other organizations are emerging that focus on areas such as equine-facilitated mental health and equine-facilitated psychotherapy.

EQUINE MANAGEMENT TERMS

American Horse Council An organization that represents the horse industry in Washington, D.C., and works to promote legislation supportive of the horse industry.

American Hippotherapy Association An organization that promotes research and professional development for people involved in hippotherapy.

Appaloosa coat patterns Several breeds show the coat patterns most associated with the Appaloosa and the POA. The following are common coat patterns seen in Appaloosa and POA horses:

Blanket Dark body with white over some parts of the loins and hips (see Figure 14-22). May be just the white blanket, or a white blanket with colored spots matching the dark body on the white background.

Leopard A white body with colored spots on the whole body (refer to Figure 14-18). Some horses are heavily spotted, whereas others have a small number of spots; those are known as **few spot leopards.**

Snowflake A dark body with white hairs over the hips and/or loin.

Bay A term describing a coat that is light brown to reddish-brown with a black mane, tail, and legs. May have white markings.

Bedouin (behd-ō-əhn) A nomadic tribe of the Arabian Desert that developed the Arabian horse as a warhorse and for transportation.

Bit A piece of metal that fits in a horse's mouth and provides control for riders when riding. The determination of curb and snaffle is based on where the bit applies pressure on the horse. Either type can have a solid or broken mouthpiece.

Curb bit A bit with a shank extending from the mouthpiece to the reins, creating leverage action when pressure is applied to the bit through the reins.

Snaffle bit A bit with reins that connect directly to a ring attached to the mouthpiece of the bit. This direct connection does not provide leverage such as that seen in using a curb bit.

Bridle A piece of equipment that riders place around a horse's head to provide control when riding (see Figure 14-22).

Buckskin A dark golden color with a black mane, tail, and legs. May have white markings.

Cart A two-wheeled driving vehicle.

Cold blood Describes horses that are primarily of draft blood, with thick skin and hair, often heavy bones, and a docile temperament.

Chaps Originally, leather coverings for the leg worn to protect riders from brush while riding. Chaps are now primarily an accessory used in Western riding disciplines.

Double register Referring to an animal registered with more than one organization (for example, a breed association and color association).

Dun A brownish golden color with the same colored or slightly darker mane. Horses often have a darker stripe down the spine, and may have striping inside the legs.

English riding Riding in a saddle without a horn, either in a forward seat, saddle seat, or dressage.

Equine (ē-kwīn) A term relating to all members of the family *Equidae,* including donkeys and asses, as well as horses and ponies.

Equestrian (ē-kwes-trē-ǝhn) Relating to horses, or to a person who works with horses.

Equestrienne (ē-kwes-trē-ehn) A female who works with horses.

Equine-assisted programs A broad term for all activities involving horses to improve the mental, emotional, or physical health of people. Equine-assisted programs range from therapy, in conjunction with a trained therapist, to visitation programs where interaction with the animal creates a feeling of well-being in the patient.

Face markings Horses have a variety of white markings on the face. The markings on the face and legs are important for animal identification. Horses may have a combination of markings (for example, a star and a snip). Common face markings include the following (see also Figure 14-23):

Bald White covering the entire forehead, over the eye sockets, and over the upper lip. Often includes the lower lip as well.

Blaze A wide, white marking beginning between the eyes and going to the muzzle.

Star Any white marking on the forehead.

Strip A narrow white stripe down the face from the forehead to the muzzle.

Snip A white marking on the upper or lower lip, or between the nostrils.

Feather Long hair below the knees and hocks that may cover the hoof in some breeds.

Flying lead change Changing from one lead to another, at a canter or gallop, while still cantering or galloping. A **simple lead change** involves changing leads by going from a canter to a trot, walk, or halt, and then starting to canter again on the other lead.

Gait A pattern of footfalls resulting in movement. There are both **natural gaits,** which the horse can instinctively perform (see Figure 14-24), and **artificial gaits,** which must be learned. A horse that is **gaited** performs gaits other than the standard walk, trot, and canter. The following are terms that refer to various gaits:

FIGURE 14–23 Face markings in horses.

Walk A flat-footed, lateral natural gait in which three feet are always on the ground (right hind, right front, left hind, left front).

Trot A medium diagonal, two-beat natural gait with a **suspension phase,** when all four feet are off the ground (right hind and left front, left hind and right front move together). The term **jog** is used to describe this gait in Western show classes.

Pace A medium lateral, two-beat natural gait with a suspension phase. The left hind and left front legs move together, then the right front and right

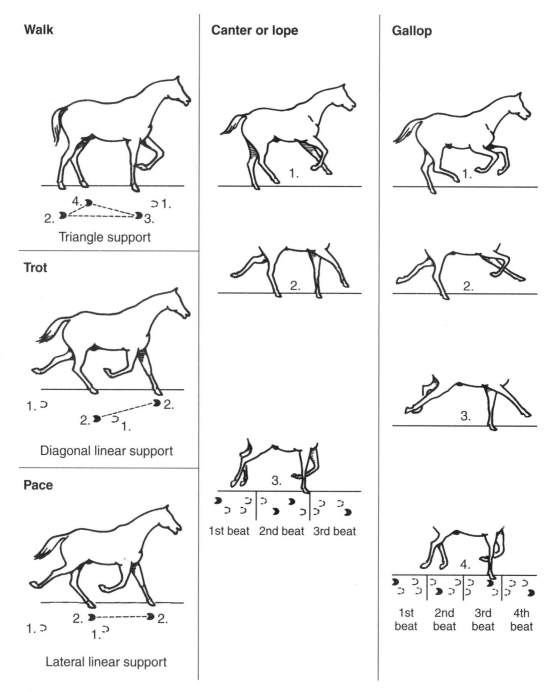

Walk

Triangle support

Trot

Diagonal linear support

Pace

Lateral linear support

Canter or lope

1.

2.

3.

1st beat 2nd beat 3rd beat

Gallop

1.

2.

3.

4.

1st beat 2nd beat 3rd beat 4th beat

FIGURE 14–24 The footfall patterns of natural gaits in horses.

hind legs. The lateral gait has a slight rolling motion from side to side. The trot is the natural gait for most horses instead of the pace. Most pacers are Standardbreds.

Running walk   The running walk is a four-beat lateral natural gait with the footfall pattern of right hind, right front, left hind, left front. The running walk differs from the regular walk in the rate of the steps, and in that the horse steps beyond the front leg with the hind leg. The running walk has no suspension phase, which makes it extremely smooth. The running walk is a breed characteristic of the Tennessee Walking Horse.

Rack   The rack is an artificial gait for most horses, and must be taught. However, some horses are predisposed to this gait, including American Saddlebreds and some other gaited breeds. The rack is a four-beat gait with the same footfall pattern as the walk and the running walk. The rack has an equal amount of time between each step. The **slow gait** is a slower version of the rack.

Canter A three-beat natural gait with a suspension phase. The horse begins with one hind leg, then the other hind leg and the diagonal front leg move together, followed by the other front leg. The canter is not a symmetrical gait. If the horse begins with the left hind leg, the right front leg is independent, and they are on the right lead; if the horse begins with the right hind leg, and the left front leg is independent, they are on the left lead. The word **lope** is used to describe this gait in Western show classes.

Gallop The fastest of the natural gaits and most similar to the canter, the gallop is a four-beat gait. The difference from the canter is that instead of the second part of the stride being one hind leg and one front leg in unison, they strike the ground separately. On the right lead: left hind, right hind, left front, right front. On the left lead: right hind, left hind, right front, left front. The suspension phase occurs between the final front leg, and the hind leg beginning the next stride.

Groom (groom) (1) To clean a horse with brushes; or (2) a person who is responsible for the care of the horse.

Grooming equipment The brushes commonly used to groom a horse (see Figure 14-25). The following are examples of grooming equipment:

Curry comb A rubber disk with blunted teeth that works dirt and hair loose. The curry comb should not be used on sensitive areas such as the face and lower legs.

Stiff brush A bristled brush with stiff bristles that is used to remove the loosened hair and dirt from the horse.

Soft brush A brush with soft bristles used to brush sensitive areas such as the face and lower legs. The soft brush is also run over the entire body following the stiff brush to remove dust.

FIGURE 14–26 A horse wearing a nylon halter. (Courtesy of Purdue University)

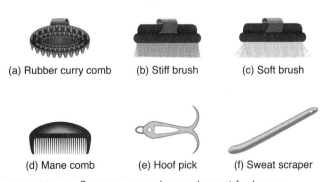

(a) Rubber curry comb (b) Stiff brush (c) Soft brush

(d) Mane comb (e) Hoof pick (f) Sweat scraper

FIGURE 14–25 Common grooming equipment for horses.

Hoof pick A hooked implement used to remove debris from the hoof of the horse. The hoof pick should always be used from the heel to the toe to ensure that any sharp debris is not pushed into the sensitive parts of the hoof.

Halter A piece of equipment used to control the horse when leading or tying. The halter goes over the poll of the horse and across the nose. The halter does not have a part in the mouth of the horse (see Figure 14-26). Halters can be made of rope, leather, or nylon.

Hand A unit of measurement for measuring the height of a horse at the withers. One hand equals 4 inches. A horse that stands 15.2 hands tall is 15 hands plus two inches.

Harness Equipment used to attach a horse to a cart or wagon for pulling.

Hippotherapy The use of the unique movement of the horse for physical benefits to a patient with the assistance of a licensed therapist.

Hot blood A term that describes horses of generally Arabian and Thoroughbred ancestry. Hot-blooded horses are very athletic light horses with thin skin and fine hair. May be associated with high energy levels.

Jockey A person that rides a racehorse.

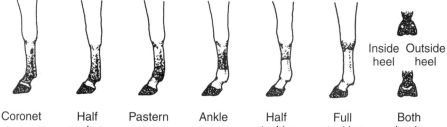

FIGURE 14–27 Common leg markings of horses.

Coronet | Half pastern | Pastern | Ankle | Half stocking | Full stocking | Both heels

Inside heel Outside heel

Leg markings White markings that begin at the hoof and goes various distances up the leg (see Figure 14-27). The following are common leg markings:

Coronet White markings just around the coronet band.

Ermine spots (er-mihn) Dark spots in a white leg marking, usually at the coronet.

Heels White on one or both heels.

Pastern White up to the bottom of the fetlock joint.

Half pastern White to the middle of the pastern.

Sock White over and including the fetlock joint.

Stocking White to the knee or hock.

North American Riding for the Handicapped Association (NARHA) An organization that certifies instructors and accredits facilities offering therapeutic horseback riding and associated activities.

Overo (ō-vār-ō) A marking pattern for Paint or pinto horses that are irregular and scattered across the body. With an overo pattern, the markings usually have uneven edges, with one or more legs often dark. The tail is usually solid in color, and the white markings do not cross over the back from one side to the other.

Palomino (pahl-ah-mē-nō) A palomino is a horse with a coat the color of a gold coin, and a white or silvery white mane and tail. The body color can range from cream through a dark gold, but the bright gold is preferred. Palominos also may have white markings. The palomino color occurs in numerous breeds, and the Palomino Horse Breeders Association will register horses of any ancestry that meet the color requirements of the registry. Many owners choose to **double register** their horses with the breed association and the palomino association.

Piebald A horse with a black-and-white patched color pattern.

Pinto (pin-tō) A coat pattern of white and any other color in large patches (see Figure 14-28). Pinto is a color and many breeds and individuals can have a pinto coat color. Paint is a breed, such as the American Paint Horse; only horses eligible for registration as American Paint Horses should be called "Paints."

Posting (pōst-ing) The act of a rider rising from the saddle and sitting down in the saddle in rhythm with the trot. Because of the suspension phase, the trot is

FIGURE 14–28 Pinto horses crossing a river. (Courtesy of ARS/USDA)

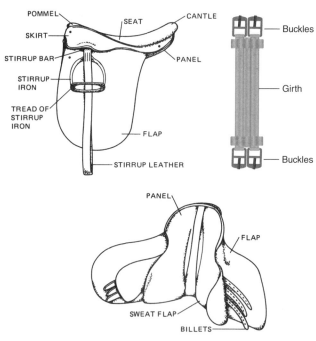

FIGURE 14–29 Parts of an English saddle. (Courtesy of University of Illinois at Urbana-Champaign)

a bouncy gait to sit, and posting protects both the horse and rider from fatigue.

Rein (rain) A strap extending from the bit to the rider's hand.

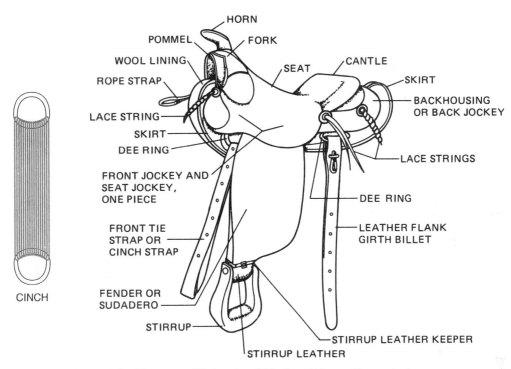

HORN
POMMEL
FORK
WOOL LINING
SEAT
CANTLE
ROPE STRAP
SKIRT
LACE STRING
BACKHOUSING
OR BACK JOCKEY
SKIRT
DEE RING
LACE STRINGS
FRONT JOCKEY AND
SEAT JOCKEY,
ONE PIECE
DEE RING
FRONT TIE
STRAP OR
CINCH STRAP
LEATHER FLANK
GIRTH BILLET
FENDER OR
SUDADERO
STIRRUP
STIRRUP LEATHER KEEPER
STIRRUP LEATHER
CINCH

FIGURE 14–30 Parts of a Western saddle. (Courtesy of University of Illinois at Urbana-Champaign)

Roan (rōn) A coat color with intermixed solid-colored hairs and white hairs. A **red or strawberry roan** is a combination of chestnut and white hairs, and **blue roan** is a combination of black and white hairs.

Saddle Equipment placed on a horse's back that provides a seat for riding. Saddles fall in two primary categories: English saddles (see Figure 14-29), which do not have a saddle horn, and Western saddles (see Figure 14-30), which have a saddle horn. Each saddle was developed for a specific purpose, and saddles are still used for traditional purposes, as well as for recreational riding.

Saddle pad A thick piece of cloth placed between the horse and the saddle that absorbs sweat and cushions the horse's back from the saddle.

Skewbald (skyew-bahld) A patched color pattern of white and any color except black.

Stock type A muscular body type that is consistent with that desired for working on a ranch. The Quarter Horse body type is a "stock type."

Stride When a horse is moving, the distance from the first step of a gait, through all subsequent steps of the gait, until the first step is repeated.

Sulky A lightweight two-wheeled cart for racing.

Tack Equipment used when riding or driving a horse.

Tobiano (tō-bē-ah-nō) The tobiano is a color pattern for Paint or pinto horses that is in large patches, instead of scattered. The ends of the pattern tend to be uniform, and white often crosses over the back.

The mane and tail reflect the pattern, and are often partly colored. The head is usually dark, with markings similar to those seen in solid-colored horses.

Tovero (tō-vēr-ō) Tovero is a color pattern for Paint or pinto horses that shows characteristics of both overo and tobiano patterns.

Warm blood A group of breeds developed from a combination of hot-blood and cold-blood ancestry. The horses are generally athletic and of good size and substance. Some breeds have now adopted the use of the term *warm blood* as a part of the name.

Western riding (1) A style of showing that includes the use of Western saddles, equipment, and attire. Riders usually wear hats, and in many classes also wear **chaps.** (2) A specific class in which a horse performs a pattern around cones and over logs.

CHAPTER SUMMARY

The horse has been a valuable part of human society for all of recorded time. Horses have served every role possible in society in the thousands of years of interaction between horses and humans. Cave paintings even show horses being hunted as food. Horses have gone from a primarily work animal, to an animal that is used for work, recreation, and therapy. There are more horses in the United States today than when horses still provided the primary source of power for humans.

STUDY QUESTIONS

Match the breed with the fact provided.

1. _____ Arabian   a. Best known for racing on the flat.

2. _____ Percheron   b. Results when a horse and donkey are bred.

3. _____ Quarter horse   c. The smallest of the pony breeds.

4. _____ Appaloosa   d. Naturally performs the running walk.

5. _____ Mule   e. The oldest and purest breed.

6. _____ Shetland   f. A golden draft horse.

7. _____ Saddlebred   g. A stock-type horse with a distinctive spotted coat.

8. _____ Tennessee Walking Horse   h. A gray or black draft horse.

9. _____ Belgian   i. The "peacock of the show ring."

10. _____ Thoroughbred   j. The most popular breed in the United States.

11. What rodeo event involves jumping off the horse to tackle a steer?
    a. Calf roping
    b. Steer wrestling
    c. Cow tipping
    d. Steerdogging

12. How many barrels are in a barrel racing pattern?
    a. 4
    b. 3
    c. 2
    d. 1

13. What three horses were the foundation sires for the Thoroughbred?

14. What organization is the legislative arm of the horse industry?

15. What color is a palomino?

16. What Indian tribe created the Appaloosa horse?

17. What breed races with a sulky?

18. What three races are in the Thoroughbred Triple Crown?

19. Draw and label three leg markings.

Chapter 15
Canine

Dogs were one of the early species that man domesticated. They have played many roles in human history. It is generally believed that dogs first lived in proximity to humans because they were attracted to the refuse of the human community as a food source. Dogs now serve in a wide variety of roles, and humans have developed a diversity of breeds unseen in other species.

BREEDS

The American Kennel Club (AKC) is the registering body for dog breeds in the United States. The AKC currently recognizes 150 breeds, and divides them into seven groups and a miscellaneous class. Most breeds were developed with a specific purpose in mind. The following is a sampling of breeds from each group; refer to the American Kennel Club (www.akc.org) for a complete list.

Hound Group

The hound group is comprised of dogs that were primarily developed for hunting. These dogs have characteristic drooping ears, and a wide range of sizes.

Scent hounds   These hounds identify prey by scent, and then track their prey until caught. Scent hounds often have a distinctive "**baying**" sound they make when their prey is cornered.

  Basset hound   Dogs with short legs, long bodies, and very substantial bone structure, Basset hounds were originally bred in France. The dogs have a short hair coat that is a combination of white and brown (see Figure 15-1). Bassets are primarily used for hunting rabbits.

  Beagle   Originally bred in England for hunting rabbits, the beagle is brown and white with a short hair coat and sturdy legs (see Figure 15-2). The beagle is an active dog and an instinctive hunter. Beagles were recognized by the AKC in 1885.

  Black-and-tan coonhound   A moderate-sized dog that is black with tan markings (see Figure 15-3).

FIGURE 15–1 Bassett Hound (Courtesy of Mary Bloom@AKC)

193

FIGURE 15–2 Beagle (Courtesy of Mary Bloom@AKC)

FIGURE 15–3 Black and Tan Coonhound (Courtesy of Mary Bloom@AKC)

FIGURE 15–4 Afghan Hound (Courtesy of Mary Bloom@AKC)

Coonhounds were developed in the United States for hunting raccoons, and are active dogs with a short hair coat.

Bloodhound One of the largest of the hound breeds, the bloodhound was developed in England from primarily French breeds. This old breed is characterized by a brown to reddish-brown coat with black over the back. They have long ears and loose skin that hangs in folds around the face.

Dachshund (dox-hund) The Dachshund was developed as far back as the 1400s in Germany for hunting small animals such as badgers. The Dachshund is short-legged and long-bodied, and can be black, brown, or red. The breed has three coat-type varieties: short-haired, long-haired, and wire-haired, as well as a miniature variety. The Dachshund was recognized by the AKC in 1885.

Sight hounds These dogs visually spot prey and then chase it down. Sight hounds are generally very fast, light-framed dogs with deep bodies.

Afghan hound This fairly large dog is characterized by a long, silky coat that can be in any color, and a face with short hair (see Figure 15-4). The Afghan originated in Afghanistan where it was used for hunting.

Basenji An ancient breed that originated in Zaire, the basenji is unique in that it does not bark like other dogs. Basenjis are not silent, as they have a yodeling vocalization. They also have a short hair coat and erect ears.

Borzoi (bōr-zoy) Originally called the Russian wolf hound, this elegant and graceful dog was the preferred dog of Russian aristocracy. They were developed for chasing and capturing wolves. These dogs can come in any color, have long hair that can be either flat or wavy, and a deep chest characteristic of many of the sight hounds.

Greyhound The greyhound is an ancient breed developed for hunting. These dogs have long, slender heads, bright dark eyes, and a short coat. Greyhounds are a variety of colors, from fawn to brindle and black. They have deep chests and tight flanks. Over a short distance, they are the fastest dogs in the world. Racing is a popular activity with greyhounds. They are great pets, and are relaxed and affectionate in the home.

Ibizan hound (ihb-ih-zəhn) The Ibizan hound is an ancient hound that was originally bred for hunting rabbits. Ibizans have large, erect ears and two coat varieties: short-haired and wire-haired. These dogs can be up to 27½ inches tall, and weigh around 50 pounds. They are white and red, either in combination, or solid. The red may range from a very light yellowish-red, through a deep red (see Figure 15-5).

Irish wolfhound A very large dog, one of the giant breeds, the Irish wolfhound is at least 32 inches tall at the shoulders. The breed was developed in Ireland before the Romans conquered Britain, and was introduced to Rome by returning soldiers. However, the breed nearly became extinct in the 1800s. The numbers are now strong,

FIGURE 15–5 Ibizan Hound (Courtesy of Mary Bloom@AKC)

FIGURE 15–6 Rhodesian Ridgeback (Courtesy of Mary Bloom@AKC)

and these large dogs with a wiry gray, black, or brindle coat are once again seen fairly commonly.

Rhodesian ridgeback (rō-dē-zhun) The Rhodesian ridgeback was developed in Africa in the 1500s, from crossing dogs that settlers brought from Europe with native dogs. Rhodesian ridgebacks were used to hunt lions, and are sometimes called African lion hounds. These dogs are fawn to red in color, with a short hair coat. The most distinctive characteristic is the swirl of hair that grows backward against the rest of the hair along the spine (see Figure 15-6).

Saluki (sa-loo-kē) The saluki is another ancient breed with a long head, long drooping ears, and long legs. Their coat is short, except on the ears and tail, which have long silky hair. Salukis range in color from white to tan, and black and tan.

Herding Group

The herding group is comprised of dogs that were bred to assist farmers in caring for their flocks and herds of animals. These animals were integral to moving and handling livestock, and many animals still serve that purpose. As a group, herding dogs tend to be high in energy, very trainable, and bond strongly to their human companions.

Australian shepherd The Australian shepherd is a medium-sized dog developed for working sheep. The dogs are extremely intelligent and are protective of their families, as well as of their sheep. They have long hair, and can be blue merle, red merle, or tricolored (see Figure 15-7). They are larger-bodied than border collies, but like border collies, they are very trainable and excel at events such as agility competitions.

Border collie The border collie is a medium-sized dog of exceptional intelligence, energy, and trainability. First bred on the southern border of Scotland, the border collie is primarily black and white, but can be a variety of colors. These dogs can either have a moderately long or a short coat (see Figure 15-8). The dogs are still used extensively for sheep herding, and are tremendously successful in sports such as agility. Border collies have very active minds and bodies, and must be kept active.

Collie The collie was developed in Scotland as a herding dog. Collies can be rough-coated with long hair, or smooth-coated with short hair. Regardless of the type of hair coat, collies are sable, tricolor, or

FIGURE 15–7 Australian Shepherd (Courtesy of Mary Bloom@AKC)

FIGURE 15–8 Border Collie (Courtesy of Mary Bloom@AKC)

blue merle, with white markings. They have long, relatively narrow heads and intelligent eyes. Collies should retain the movement and athleticism consistent with their heritage as herding dogs. The famous movie dog Lassie was a collie.

German shepherd   The German shepherd is a very popular dog in the United States. Sometimes, popularity leads to inattentive breeding and the development of undesirable characteristics in the dog. The well-bred German shepherd is an excellent dog that is intelligent and possesses a sound and stable temperament. These dogs are black, or black and tan, and rarely white. They have a moderately wide head, erect ears, and a substantial body. However, some care should be taken to ensure that German shepherds are purchased from a reputable breeder. Poorly bred dogs have a higher incidence of hip dysplasia and less stable temperaments.

Old English sheepdog   Developed in western England for moving sheep and cattle to market, the Old English sheepdog is a moderately large dog that is gray, blue, or blue merle, and may or may not have white markings. These dogs have an extremely long and thick coat that is shaggy and relatively straight. They have a relatively square head and are quite good-natured. Extensive grooming is required to keep their long hair from becoming matted.

Puli (poo-lē)   Pulis were developed in Hungary to herd sheep. They are medium-sized with a distinctive corded coat that resembles dreadlocks. They can be black, white, gray, or apricot (see Figure 15-9). Because of their distinctive coats, pulis require extensive grooming.

Welsh corgi (kohr-ghē)   The Welsh corgi was developed as a cattle dog in the 1100s. These dogs are very active, intelligent dogs with short legs, relatively long and powerful bodies, and deep chests. There are two breeds of Welsh corgi: the Pembroke, which is a shade of brown from fawn through sable, or black, with white markings and a docked tail; and the Cardigan, which can be of any color and has a full-length tail. Both breeds have erect ears.

Nonsporting Group

The nonsporting group has a wide range of dogs that do not easily fit into the other groups. Nonsporting dogs have a tremendous range in size, hair coats, uses, and temperaments.

American Eskimo dog   A white dog with erect ears, and a pointed face. The American Eskimo dog has a thick undercoat and a long straight outercoat. In this breed, the tail curls over its back. American Eskimos have three size categories: toy (9–12 inches), miniature (12–15 inches), and standard (more than 15 inches).

Boston terrier   The Boston terrier was developed from crossing English bulldogs with native terriers in Boston. They are smooth-coated, with a short head and large round eyes. Their ears are erect, and they are deep-bodied with wide chests (see Figure 15-10). Boston terriers are small dogs, and are used primarily as companion dogs.

Chinese shar-pei (shar pay)   The shar-pei originated in China over 2,000 years ago as a fighting dog. As recently as the 1980s, the dog was nearly extinct; however, focused breeding has increased the numbers of this breed. The dogs are black, red, fawn, or cream, with distinctive deep wrinkles in the skin around the face (see Figure 15-11). Because the

FIGURE 15–10 Boston Terrier (Courtesy of Mary Bloom@AKC)

FIGURE 15–9 Puli (Courtesy of Mary Bloom@AKC)

FIGURE 15–11 Chinese Shar-Pei (Courtesy of Mary Bloom@AKC)

modern shar-pei is descended from a small genetic pool, it is very important to ensure that the dog is not carrying genetic defects.

Dalmatian Believed to have been originally developed in the region of Dalmatia on the Adriatic Sea, the Dalmatian is a medium to large dog, with a white coat and black or liver spots. Their hair is short and smooth, and their ears droop. Dalmatians are well known as carriage dogs, and for being firehouse mascots. Dalmatians have a genetic predisposition for deafness in some lines, so the hearing of puppies should be tested.

Poodle The poodle was first developed in France where it was used as a hunting dog. The poodle has a long head, almond-shaped eyes, and drooping ears. They come in a wide range of colors. Poodles range in size from the standard poodle, which is more than 15 inches tall, to the toy poodle, which is in the toy group. Their hair coat is unique and curly, does not shed as much as other dogs, and is less likely to cause allergic reactions.

Schipperke (skihp-er-key) A small black dog developed in the Netherlands, the schipperke name means "little captain." The schipperke has relatively long hair that is thick and dense, and stands up around the neck (see Figure 15-12). These dogs worked as watchdogs on barges moving through the channels of the Netherlands. Their tails are often docked, but curl over the back when left natural.

Sporting Group

Dogs in the sporting group were developed for hunting. Retrievers, pointers, and spaniels are all in the sporting group. The Labrador retriever, the most popular breed in the United States according to the AKC, is a member of the sporting group. The following are members of this group:

Cocker spaniel The cocker spaniel originated in England. There are two varieties, the American cocker spaniel and the English cocker spaniel. Both varieties are relatively small, with long silky hair, and liberal

FIGURE 15–12 Schipperke (Courtesy of Mary Bloom@AKC)

FIGURE 15–13 Cocker Spaniel (Courtesy of Mary Bloom@AKC)

FIGURE 15–14 German Short Haired Pointer (Courtesy of Mary Bloom@AKC)

feathering on both the front and hind legs. Cockers have long drooping ears that are also liberally feathered (see Figure 15-13). These dogs come in a wide range of colors.

English setter The English setter is a medium-sized dog with long wavy hair, which creates feathers on the front legs. English setters have a long head, and drooping ears that are covered with moderately long hair. They come in a variety of colors intermixed with white. The English setter identifies game by pointing.

German shorthaired pointer A medium-sized dog with a short hair coat that ranges from liver to black, and may be combined with white (see Figure 15-14). German shorthaired pointers are strong, good-natured dogs that are willing to work.

Golden retriever A moderately large dog with long hair that can be wavy or flat, the golden retriever ranges from cream through a deep gold color. Golden retrievers are substantial dogs and are known for their excellent temperaments. They are regularly one of the most popular breeds in the United States.

Labrador retriever An excellent breed of retriever that was originally developed in the Newfoundland province of Canada, not Labrador as the name would imply. The Labrador is black, yellow, or chocolate, with a short, thick hair coat. Labs are well muscled,

FIGURE 15–15 Labrador Retriever (Courtesy of Mary Bloom@AKC)

FIGURE 15–16 Airedale (Courtesy of Mary Bloom@AKC)

FIGURE 15–17 Parson Russell Terrier (Courtesy of Mary Bloom@AKC)

strongly built dogs, with a deep, wide chest. Labradors have a distinctive "**otter tail**," which is wide at the base, and narrows to a point (see Figure 15-15). The Labrador has been the most popular registered breed in the United States for several years.

Vizsla (vēsh-lah) The Vizsla is also called the Hungarian pointer, and was developed in Hungary. Artwork from the 1300s shows vizsla-like dogs involved in hunting. These dogs have a short, tight coat that is a golden red in color, a broad forehead, and long, thin, drooping ears. The tail of the vizsla is often docked.

Terrier Group

Terriers (tār-ē-ərs) are a group of relatively small dogs, with the exception of the Airedale terrier. These dogs were developed for chasing prey, primarily vermin, to ground. They often went into dens to retrieve animals. Terriers are typically very active, intelligent dogs, with independent personalities.

Airedale terrier The Airedale is the largest of the terriers, and can grow up to 24 inches tall. The Airedale is primarily tan, with a black saddle over the back. Their hair coat is wiry, and they have extensive whiskering around the muzzle (see Figure 15-16).

Cairn terrier (kār-n) The cairn terrier was first bred in Scotland near the Isle of Skye, and probably shares some ancestry with the Skye terrier. Cairn terriers can be any color except white. They have a thick outer coat and a dense undercoat. Cairns are small, and are generally less than 12 inches tall.

Parson Russell terrier A bright active dog that is also known as the Jack Russell terrier, the Parson Russell was first bred in England by a minister who was seeking a smaller version of the fox terrier. These dogs are white, with brown markings on their head and hindquarters. They come in either smooth-coated or wire-coated varieties (see Figure 15-17).

Miniature Schnauzer (shnow-zer) The Schnauzer is of German descent, and varies in color from black to salt and pepper, which is a blend of black and white. These dogs have short hair over the body, with longer hair on the legs, and generous whiskers on the face. The standard schnauzer is a larger version of the miniature schnauzer and is a member of the working group.

Scottish terrier An old breed of undetermined origin, the Scottish terrier or Scottie is a small dog (less than 12 inches), with a relatively large head, and a distinctive expression as a result of the prominent eyebrows and the generous whiskering of the muzzle (see Figure 15-18). These dogs have liberal feathering on the legs, and can be black, wheaten, or brindle.

Toy Group

Dogs in the toy group are small, and have always been bred as companions or "lap dogs." The dogs in this group originate from all parts of the world. The following are some of the dogs in the toy group:

Chihuahua (chih-wah-wah) The Chihuahua is known as the world's smallest dog, and weighs less than 6 pounds. They are named after the Chihuahua region of Mexico, although whether the breed was developed there or brought to that region is not clear. These dogs can be either smooth-coated, or long-haired, with a round head and round eyes (see Figure 15-19). Chihuahuas were developed as companion animals, and that is still their major role.

FIGURE 15–18 Scottish Terrier (Courtesy of Mary Bloom@AKC)

FIGURE 15–19 Chihuahua (Courtesy of Mary Bloom@AKC)

FIGURE 15–20 Pekingese (Courtesy of Mary Bloom@AKC)

FIGURE 15–21 Shih-Tzu (Courtesy of Mary Bloom@AKC)

Papillon (pahp-ē-yōn)   The *papillon,* which is French for butterfly, was so named because of the characteristic butterfly shape of the ears. Their head is rounded with round eyes, their hair is long and silky, and they have white fur with brown or black markings.

Pekingese (pēk-eh-nēs)   The Pekingese is an ancient dog developed in China, and prized by Chinese royalty from at least the eighth century. The "Peke" is a small dog with a broad head and a short face. These dogs have long straight hair with considerable feathering on the legs and around the neck (see Figure 15-20). Almost any color is allowed for registration.

Pomeranian (pohm-ər-ān-ē-ahn)   The Pomeranian was originally bred as a herding dog, but has been reduced in size over the last century to the 5-pound dog we know today. The "Pom" can be any color, with a long outercoat and a short dense undercoat. Their tail curls over their back, and "Poms" have sharp, erect ears and a sharp-featured face. They are happy, alert dogs, and are popular as companions.

Shih Tzu (shē tsoo)   The Shih Tzu is an ancient Asian dog of unknown breed origin. The earliest records show the Shih Tzu as a member of the Chinese royal court. These bright and alert dogs have a long thick coat, large dark eyes, and a strong body. Their tail curls over their back and is covered in long hair (see Figure 15-21). These dogs can be any color.

Yorkshire terrier   The "Yorkie" was originally bred in northern England for hunting rats. The Yorkie has long, steel-gray hair that is silky and smooth, with tan markings. Their ears are erect, their eyes are bright, and they have the typical terrier "large dog in a little body" temperament.

Working Group

The working group is made of a variety of dogs that are involved in some type of work. This may include dogs used for draft work, protecting herds of sheep or goats, or protecting people and property. The following are dogs classified in the working group:

Boxer   The boxer is an active dog that was originally developed in Germany as a guard dog. These dogs are of good size, with a square head and short muzzle. They have loose lips, with wrinkles around the face. Boxers have naturally drooping ears, but some breeders choose to crop the ears. Dogs can be fawn or brindle, and may have white markings (see Figure 15-22).

FIGURE 15–22 Boxer (Courtesy of Mary Bloom@AKC)

FIGURE 15–23 Great Pyrenees (Courtesy of Mary Bloom@AKC)

Doberman pinscher (pin-cher)   The Doberman is a moderately large dog that was developed in the 1860s by Louis Dobermann. These dogs are black or red, with a short, smooth coat. Their heads are long and lean with bright, dark eyes. This bright and active dog should not be aggressive.

Great Dane   Great Danes are very large dogs, and originated in Germany, where they were bred for hunting boars. These dogs have broad heads, drooping ears, and are very muscular. Their hair coat is short and thick, and can be brindle, fawn, blue, black, or harlequin.

Great Pyrenees (pēr-ah-nēz)   A large white dog with long hair and a thick undercoat (see Figure 15-23), the Great Pyrenees was developed in the Pyrenees mountains between France and Spain as a guard dog to protect sheep. Great Pyrenees still serve in that role in sheep herds in the United States. They live with the sheep full-time, and protect them from predators. These dogs are very docile toward people, and need extensive grooming to maintain their long hair when kept in close contact with people.

Newfoundland (new-fund-land)   The Newfoundland is a massive dog that was developed in Canada as a draft dog, for pulling loads. These dogs have a broad head, drooping ears, and a long thick coat. They can be black, brown, or a combination of black and white. These intelligent dogs are excellent companions.

FIGURE 15–24 Siberian Husky (Courtesy of Mary Bloom@AKC)

Rottweiler (rhot-wī-ler)   The Rottweiler is a large black and tan dog developed in Germany as a herding and guard dog. These dogs have a broad head and broad chest, and are very muscular. Rottweilers may have a reputation for aggression, but dogs from good breeding programs are intelligent and have excellent temperaments.

Siberian husky   The Siberian husky was developed as a sled dog by the native Chukchi people of northern Asia. This medium-sized dog has a long, furry coat that keeps it warm in the cold Arctic. These dogs are black or gray with white markings, or may be all white (see Figure 15-24).

Miscellaneous Group

The miscellaneous group is made of dog breeds that are on the "waiting list" to be accepted into the American Kennel Club. The breeds in this group vary depending on the year. After receiving full approval from the AKC, the breed will be assigned to the appropriate permanent group. For a list of the current breeds in the miscellaneous group, check the AKC Web Site.

USES

The following are uses for dogs:

Companions   Most dogs are companions or pets. In many families, the dog is a part of the family, and the emotional tie is very strong.

Service   Service dogs assist people in conducting day-to-day tasks. Service dogs do tasks that range from leading people who are blind, to helping people in wheelchairs turn lights off and on. Service dogs require extensive training, and many organizations such as Canine Companions for Independence, the Delta Society, and Leader Dogs for the Blind train dogs and facilitate their placement with people with disabilities.

Sport

People participate with their dogs in a wide variety of competitive activities. Many of these sports are based in activities that the dogs were developed to accomplish. The following are different sports and activities in which people participate with their dogs:

Agility   A competitive activity in which dogs race over an obstacle course.

Bench   Describes dogs that are exhibited and judged based on their conformation and breed type.

Field trial   A competitive activity in which the ability and skills of hunting dogs and hounds are evaluated.

Fly ball   A competitive event in which dogs run down a track, release and catch a ball, and run back to the start line.

Lure coursing   A competition in which dogs, primarily sight hounds, chase a lure over a predetermined course. Dogs are scored on speed, endurance, agility, and how well they follow the lure.

Obedience trial   A competition in which dogs and handlers are evaluated on the dog's ability to follow specific commands from the handler. To earn AKC certificates or points, the AKC must license the trial. The following are certificates awarded for obedience:

CD   Companion Dog Certificate awarded by the AKC to qualifying dogs in the novice obedience class.

CDX   Companion Dog Excellent Certificate awarded by the AKC to qualifying dogs in the open obedience class.

UD   Utility Dog Certificate awarded by the AKC to qualifying dogs in obedience Utility class.

Tracking   Tracking is a competition that has been developed around the skills and partnership of a dog and handler necessary for successful search and rescue work. The following certifications are available through the AKC for dog/human teams that are successful trackers:

TD   Tracking Dog. A title for a dog that has successfully followed a track that a person has laid along a course 440–500 yards in length. The scent must be laid 30 minutes to 2 hours prior to the beginning of the exercise. Three to five changes of direction are required in a track that meets the requirements for the TD title.

TDX   Tracking Dog Excellent. A dog must successfully follow a longer track (800–1,000 yards) that is also older (three to five hours) than the track required for the TD title. The course also has five to seven changes of direction, and will have additional human scent, besides the one the dog is supposed to be tracking, to increase the challenge for the dog.

FIGURE 15–25   A German Shepherd dog in training as a police dog.

Work

Many dogs still work for their owners. They may also have a role as companions in the family. The following are working roles that dogs perform:

Contraband detection   Dogs are trained to identify the scents of illegal drugs, food items, or other things that may be brought into the country or transported. Dogs are taught a behavior to indicate when they have found something suspicious. This **marking behavior** can be barking, sitting, or any other behavior that informs the handler that the dog has found something.

Herding   Primarily sheep and cattle herding. Herding dogs assist in gathering and moving animals from one place to another, often into corrals or pens. For example, the border collie is used primarily to herd sheep, and the corgi is used for cattle.

Police dogs   Police dogs are partners in law enforcement. They assist with detection and apprehension of suspects, and protect their human partners (see Figure 15-25).

Search and rescue   Dogs are used extensively to find and rescue people who have become lost.

CANINE MANAGEMENT TERMS

American Boarding Kennel Association   An organization that certifies boarding kennels.

American Kennel Club (AKC)   The AKC is the largest dog registering organization in the world, and was established in 1884. The AKC sponsors many of the most important dog shows in the United States.

Animal shelter A place for animals that do not have a home. Animals may be placed in shelters by their owners, or they are found as strays. Most animals in shelters are euthanized due to lack of space.

Association of American Feed Control Officials (AAFCO) An organization that oversees the animal feed industry, and ensures that regulations governing the industry are followed.

Bitch A female dog.

Blue merle (murl) The color of blue and gray intermixed with black (refer to Figure 15-7).

Boarding kennel A facility that cares for other people's animals on a short-term basis.

Body condition The amount of fat on an animal evaluated on a scale of 1–10. A score of 1 indicates an animal that is extremely emaciated (see Figure 15-26). A score of 5 is assigned to a dog that is in good body condition (see Figure 15-27). A score of 10 indicates a dog that is grossly obese (see Figure 15-28). Both weight extremes have health problems associated with them.

Breed type Characteristics that identify a breed as unique from other dogs.

Brindle A mixture of black and other colored hair. The black is uniformly mixed throughout the hair coat.

Champion A designation a dog earns by meeting AKC qualifying requirements. Champion (or the abbreviation "Ch.") permanently becomes part of the dog's name.

Choke chain A chain collar that operates like a noose. The tighter the leash is pulled, the tighter the collar becomes around the dog's neck. It is important to adjust the choke collar properly, or it will not loosen when pressure is released.

Collar A fabric, nylon, or leather strap that fits around a dog's neck to provide control.

FIGURE 15–26 Body condition score 1. Hills drawing

FIGURE 15–27 Body condition score 5. Hills drawing

FIGURE 15–28 Body condition score 10. Hills drawing

Cynology (si-nohl-o-gē) The study of canines.

Digest A flavoring ingredient in pet foods made from digesting animal tissue with enzymes.

Dispostion Temperament.

Docking The cutting off of all or a portion of the tail. The amount of the tail left varies with breed.

Doggy day care A facility that provides care for a dog during the day, while the owner is at work. Many doggy day cares also offer boarding.

Ear cropping The surgical removal of some part of the outer portion of an ear. This practice is controversial, and has been banned in the United Kingdom. Some veterinarians in the United States refuse to crop ears for cosmetic purposes.

Euthanasia (yoo-thah-nā-zha) The humane killing of an animal. Millions of dogs and cats are euthanized in animal shelters annually.

Fancier (fan-see-er)   A person who is actively involved in breeding purebred dogs or cats.

Feather   Long hair that may occur on the ears, body, or legs.

Field champion   A dog that competes in field trials and has met the AKC qualifications for being a champion. Designated by **Field Champion** (or the abbreviation "Field Ch.").

Gait   The movement of an animal.

Grooming (groom-ing)   The brushing and otherwise cleaning and neatening the hair coat and nails of a dog.

Groomer   A person who grooms dogs. Grooming often includes clipping toenails and cutting hair.

Guaranteed analysis   On the bag of commercial dog food, the guaranteed analysis provides information on the nutrients in the food (see Figure 15-29).

100% Complete and Balanced

Holistic Select® Lamb Meal & Rice Formula provides complete and balanced nutrition for maintenance and is comparable in nutritional adequacy to a product which has been substantiated using AAFCO feeding tests.

Ingredients: Lamb Meal, Ground Brown Rice, Oatmeal, Chicken Fat (Preserved with Natural Mixed Tocopherols and Citric Acid), Dried Beet Pulp, Flaxseed, Brewers Dried Yeast, Rice Bran, Dried Egg Product, Salt, Potassium Chloride, Air Dried Peas, Ground Carrot Cubes, Inulin, Glucosamine Hydrochloride, DL-Methionine, Vitamin A Acetate, Vitamin D3 Supplement, Vitamin E Supplement, Riboflavin Supplement, Vitamin B12 Supplement, d-Pantothenic Acid, Niacin Supplement, Choline Chloride, Pyridoxine Hydrochloride, Thiamine Mononitrate, Folic Acid, Ascorbic Acid, Biolin, Rosemary Extract, Inositol, Dehydrated Kelp, Polysaccharide Complexes of Zinc, Iron, Manganese, Copper and Cobalt, Potassium Iodate, Sodium Selenite, Yucca Schidigera Extract, Lactobacillus acidophilus, Lactobacillus casei, Enterococcus faecium, B. Subtillus, Bacillus lichenformis, Bacillus coagulins, Aspergillus oryzae and Aspergillus niger.

Ingredients: Farine d' agneau, riz brun moulu, flacon d' avoine, gras de poulet (conservateurs: mélange de tocophérols et acide citrique), pulpe de betterave déshydraté, lin, levure de biére déshydraté son de riz, oeuf entier déshydraté, sel, chlorure de potassium, pois sèche à l'air, cube de carottes broyées, inuline, hydrochlorure de glucosamine, DL méthionine, acélate de vitamine A, supplement de vitamine D3, supplément de vitamine E, supplément de riboflavine, supplément de vitamin B12, d-pantothenic acide, supplément de niacine, chlorure de choline, chlohydrate de pyridoxine, mononitrate de thiamine, acide folique, acide ascorbique, biotine, extrait de romarin, inositol, varech déshydraté, polysaccharide complexe de zinc, fer, manganèse, cuivre et cobalt, potassium iodate, sodium sélénite, extrait de yucca schidigera, lactobacilles acidophiles, lactobacilles casai, faecium d'entérocoque, subtillus-b, bacilles de lichenformis, bacilles de coagulins, d'oryzae et Niger d'aspergille.

Guaranteed Analysis / Analyse Garantie:

Crude Protein / Protéines brutes	Min.	22%
Crude Fat / Matières grasses brutes	Min.	15%
Crude Fiber / Fibres brutes	Mix.	4.0%
Moisture / Humidité	Mix.	10%
Calcium / Calcium	Min.	1.3%
Phosphorus / Phosphore	Min.	0.85%
Vitamin A / Vitamine A	Min.	22,000 IU/kg.
Vitamin E / Vitamine E	Min.	125 IU/kg.
Omega 6 Fatty Acids* / Oméga 6*	Min.	1.6%
Omega 3 Fatty Acids* / Oméga 3*	Min.	0.17%
Glucosamine Hydrochloride*	Min.	400 ppm

*Total Lactic Acid Producing Live Microorganisms
  (Lactobacillus acidophilus, Lactobacillus casei, Enterococcus faecium)   240 million CFU/LB.
*Total Bacillus Organisms
  (Bacillus Subtillus, Bacillus lichenformis, Bacillus coagulins)   7 million CFU/LB.
*Protease (from Aspergillus oryzae and Aspergillus niger)[1]   280 HUT/LB.
*Cellulase (from Aspergillus oryzae and Aspergillus niger)[2]   100 Cellulase Units/LB.
*Amylase (from Aspergillus oryzae and Aspergillus niger)[3]   5 Dextrin Units/LB.

*Not recognized as an essential nutrient by the AAFCO Dog Food Nutrient Profiles.
*Ne figure pas à la liste des éléments nutritifs essentiels reconnus par l'AAFCO en matière d'alimentation pour chiens. Enzyme Functionality Statement: Product contains enzymes. Protease for protein hydrolysis; Cellulase for cellulose hydrolysis; Amylase for starch hydrolysis.

(1) One HUT unit of proteolytic (protease) activity is defined as that amount of enzyme that produces, in one minute under the specified conditions (40 deg Celsius, pH 4.7), a hydrolysate whose absorbance at 275 nm is the same as a solution containing 1.10 ug per ml of tyrosine in 0.006 N hydrochloric acid.
(2) One Cellulase Unit (CU) is that activity that will produce a relative fluidity change of one in 5 minutes in a defined carboxymethylcellulose substrate under the conditions of an assay (40 deg Celsius, pH 4.5.).
(3) One Dextrinizing Unit (DU), or (SKB), unit of alpha-amylase activity as defined as that amount of enzyme that will dextrinize soluble starch...at the rate of 1 g per hour at 30 deg Celsius.

FIGURE 15–29 Commercial pet food label.

FIGURE 15–30 A properly fitted harness

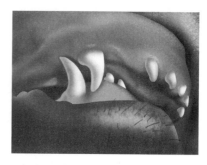

FIGURE 15–32 A dog with an overshot jaw.

FIGURE 15–31 A head halter.

FIGURE 15–33 A dog on point

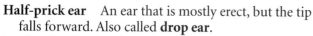

Half-prick ear   An ear that is mostly erect, but the tip falls forward. Also called **drop ear.**

Harlequin (har-leh-kin)   A color pattern of black or blue with white patches in Great Danes.

Harness   A series of straps that fit around the shoulders of a dog to provide control (see Figure 15-30).

Head halter   A nylon piece of equipment that fits around the muzzle and behind the ears of a dog. The head halter gives control of the dog, and is an excellent piece of equipment, especially for large dogs that may be difficult to control (see Figure 15-31). The head halter does not apply pressure to the throat like a choke chain, and is considered more humane, especially for nonprofessional handlers.

Kennel   A facility where dogs are housed.

Leash   A strap connecting a dog's collar to the handler's hand.

Liver   A deep reddish-brown color.

Mask   A darker color in the face, especially around the eyes.

Milk teeth   The first temporary teeth, which fall out when the permanent teeth come in.

Net quantity statement   The amount of food present in a feed container. The U.S. Food and Drug Administration (FDA) requires that the net quantity statement is on the package.

Nutritional adequacy statement   The statement on a feed package indicating whether the feed meets nutritional requirements from a feed trial using AAFCO nutrient profiles for the life stage indicated.

Obedience Trial Champion (OTCH)   A dog that has met the AKC regulations for being an Obedience Trial Champion. OTCH becomes a permanent part of the dog's name.

Open class   A class in which all dogs can compete.

Overshot jaw   When the incisors of the top jaw are farther forward than the incisors of the lower jaw (see Figure 15-32).

Pack   A group of dogs.

Pet sitter   A person who comes to a dog owner's home to care for the dog in the owner's absence.

Pet quality   An animal that shows the characteristics of a breed, but lacks the characteristics to be sold as a show or breeding animal. Some breeders require that purchasers of pet-quality animals have them spayed or neutered.

Prick ear   An erect ear.

Point   The pose of a hunting dog showing where game is present (see Figure 15-33).

Puppy   A baby dog.

Puppy mill   A facility that breeds and raises puppies in poor conditions and makes breeding decisions with little or no concern for breed improvement. Note: Not all large volume breeders are puppy mills. The term specifically refers to those who do so in poor conditions.

FIGURE 15–34 A sptiz type dog

Topknot   A tuft of hair on top of the head.
Tricolor   A coat color that has clearly defined black, tan, and white portions.
Undercoat   The fine hair that is closest to the skin.
Undershot jaw   When the teeth of the lower jaw are in front of the teeth of the upper jaw.
United Kennel Club (UKC)   The UKC was founded in 1898, and is the second largest dog registering organization in the world. The UKC is based in the United States, and emphasizes working and performance dogs.
Wire coat   A hair coat that has stiff hairs.

Red merle   A combination of red patches on a lighter background. This is similar to blue merle, with red as the base color instead of black.
Ruff   Hair around the neck that in some breeds is brushed up to frame the face.
Sable   A coat color with black hairs over a lighter base color.
Spaying   The surgical sterilization of a female.
Spitz   A type of dog usually from the Arctic region characterized by a pointed face, erect ears, and curling tail (see Figure 15-34).
Stud   A male dog used for breeding.
Tie   When dogs mate, the penis enlarges and cannot be removed from the vagina until it returns to normal size.

CHAPTER SUMMARY

Dogs have been a vital part of human life throughout history. Their original roles were focused around different types of work that dogs did to improve human lives. These vast numbers of activities and uses that dogs were involved in resulted in the formation of a wide variety of breeds. Many regions developed their own breeds for activities such as herding, draft work, and hunting. Most dogs now serve companion-type roles in our society. However, many dogs are still used as service dogs, working dogs, and in their traditional roles of hunting and herding. Competitive activities for dog enthusiasts have evolved based on many of the work roles dogs used to play.

STUDY QUESTIONS

Match the breed of dog with the group it belongs in.

1. _____ Ibizan
2. _____ Great Pyrenees
3. _____ Shih Tzu
4. _____ Dalmatian
5. _____ English setter
6. _____ Airedale terrier
7. _____ Australian shepherd

a. Terrier
b. Herding
c. Hound
d. Working
e. Toy
f. Sporting
g. Nonsporting

8. What herding breed has long curly hair, similar to dreadlocks?
   a. Puli
   b. Vizsla
   c. German shepherd
   d. Greyhound

9. What sport involves dogs running over a course of jumps and obstacles?
    a. Flyball
    b. Lure coursing
    c. Bench trial
    d. Agility

10. What are the two breeds of corgis?

11. What is the difference between a sight hound and a scent hound?

12. List three organizations that train and place service dogs.

13. What organization certifies boarding kennels?

14. What is an undershot jaw?

15. What is the miscellaneous group of dogs? Go to the Internet and find what breeds are currently in the miscellaneous group.

Chapter 16
Feline

Cats have been part of human households almost as long as dogs. Vital differences in the personalities of cats and dogs have resulted in the evolution of very different relationship between people and these two very popular companion species. A dog's willingness to be trained and work resulted in the development of a wide number of breeds suited to do a variety of tasks. The cat's independent nature made it less suited to being a working animal. The task that most cats serve is as a hunter of vermin. In fact, many theorize that cats first came into close contact with human populations because of the ready source of mice and rats that feasted on stored grains.

BREEDS

The Cat Fanciers Association (CFA) is the registering body for cats in the United States, and the largest cat registering body in the world. The CFA recognizes 41 breeds of cats. Most cats in the United States are not registered, but are mixed breed cats. Following is a sampling of breeds recognized by the CFA.

FIGURE 16–1 Abyssinian (© Larry Johnson Photography)

Abyssinian (ahb-ih-sihn-ē-ehn) The Abyssinian is an ancient breed that was first developed in Egypt around the time of the pharaohs. Much ancient Egyptian art depicts cats that are similar in type to the Abyssinian. They are medium-sized cats with large ears and large eyes, and slender yet muscular bodies. Abyssinians usually have a ticked coat pattern, and may be shades of brown, red, or blue. They are active and intelligent cats, and much beloved by their owners (see Figure 16-1).

American Shorthair The American Shorthair evolved from native cats of the eastern United States. American Shorthairs are excellent hunters and jumpers, and affectionate family pets. They may be almost any color, and have a muscular, athletic body. American Shorthairs are very popular as family pets, but are a distinct breed, not a term for any short-haired cat (see Figure 16-2).

FIGURE 16–2 American Shorthair (© Larry Johnson Photography)

FIGURE 16–3 Birman (© Larry Johnson Photography)

FIGURE 16–4 Egyptian Mau (© Larry Johnson Photography)

FIGURE 16–5 Exotic (© Larry Johnson Photography)

Birman (ber-mihn)   The Birman is another old breed from the region of Burma in Southeast Asia. This is a long-haired cat with silky hair, colorpoint markings, and distinctive white feet. These cats have blue eyes and may be a variety of colors (see Figure 16-3). They are docile cats with quiet personalities.

Burmese (ber-mēz)   The Burmese was developed in the 1930s in the United States from a cat brought home by a naval officer stationed in Southeast Asia. The Siamese was significantly used in the development of the Burmese, but the Burmese has a stockier body than the Siamese. The original Burmese were brown, but the CFA also recognizes blue, champagne, and platinum varieties. Their eyes are golden yellow in color. The Burmese has a pleasant temperament and is a very athletic cat.

Egyptian Mau (ma-ow)   The Egyptian Mau is an ancient breed that is depicted in wall paintings from ancient Egypt. The Mau has a spotted hair coat, with a distinct contrast between the color of the base coat and the color of the spots. The Mau is a strong-bodied cat, with a long body. The Mau head is a rounded wedge shape, with large ears set widely on the head. Their almond-shaped eyes are either green or gold in color. Their coat is short and dense, and may be silver, bronze, smoke, or pewter (see Figure 16-4).

Exotic   The Exotic was derived from the Persian, and shares many body characteristics with the Persian. The primary difference is that the Exotic has a short plush coat, instead of the long silky coat of the Persian. The shorter coat comes from interbreeding the Persian with breeds such as the American Shorthair, British Shorthair, Burmese, and Russian blue. Like Persians, Exotics can be a wide range of colors. They provide an opportunity for an owner to have the look of a Persian, without the labor of the long Persian hair coat to maintain (see Figure 16-5).

Maine Coon   This very popular cat is known first for its large size and second for its excellent, easy-going personality. Maine Coons can weigh up to 20 pounds, and have medium-long hair with a bushy tail, and hair on the tips of their ears. The cat originated in the eastern United States, probably from interbreeding Persians with native domestic cats. These cats come in a wide variety of colors and markings, including white markings on the face or feet (see Figure 16-6).

FIGURE 16–6 Maine Coon (Courtesy of CFA)

FIGURE 16–7 Manx (Courtesy of CFA)

FIGURE 16–8 Ocicat (Courtesy of CFA)

FIGURE 16–9 Persian (Courtesy of CFA)

Manx (mangcks) The Manx cat was developed on the Isle of Man off the coast of England. The Manx is distinguished by its lack of a tail. Manx cats are difficult to breed because mating two cats with the tailless gene results in embryonic death of the fetuses. A litter of kittens from two Manx parents can range from no tail to a full tail. Manx cats have a medium-sized body, a round head, and a thick short coat (see Figure 16-7). All cats without a tail are not of the Manx breed. The mutation causing the lack of a tail may occur in other breeds or in mixed breed cats.

Ocicat (oss-ih-cat) The Ocicat was developed in the 1960s in the United States by breeding Abyssinians, Siamese, and American Shorthairs. The Ocicat is a domestic cat, and was not developed from breeding domestic cats to the wild ocelot. The cat breed was named Ocicat because of the distinctive spots reminiscent of its wild cousin. The Ocicat is a medium to large cat with a muscular, athletic body. Their coat is short and dense, with evenly spaced spots throughout the hair coat. A variety of colors are accepted, and the spots must be clearly discernible, not blended with the base color (see Figure 16-8).

Oriental The Oriental was developed by crossing a Siamese with native cats. Orientals have the long, lean muscular body of the Siamese, but come in a wide variety of color patterns and solid colors. Orientals can be either long- or short-haired, and eye color is varied.

Persian (pər-zhuhn) The Persian was first introduced to Europe from Persia in the 1600s, and the cats were often given as gifts to royalty and other important people. The Persian is a long-haired cat with a blocky body, a round face, round eyes, and a short nose. The Persian comes in a wide variety of colors (see Figure 16-9). With a very docile temperament, the Persian ranks first in the United States in number of cats registered with the CFA.

Rex breeds Two Rex breeds are recognized by the CFA: the **Devon** Rex (see Figure 16-10) and the **Cornish Rex** (see Figure 16-11). Although two distinct breeds, both breeds are characterized by a curly or wavy coat and curly whiskers. The Devon Rex has large ears and a medium-sized body. The coarse guard hairs and the finer hairs closer to the body are both curled. The Cornish Rex is a finer-boned, more elegantly conformed cat than the Devon Rex, and does not have guard hairs, giving the Cornish Rex a softer, finer coat. The Rex breeds both developed from unplanned matings resulting in an individual with the curly coat mutation.

FIGURE 16–10 Devon Rex (Courtesy of CFA)

FIGURE 16–11 Cornish Rex (Courtesy of CFA)

FIGURE 16–12 Scottish Fold (Courtesy of CFA)

FIGURE 16–13 Siamese (Courtesy of CFA)

Scottish Fold The Scottish Fold breed descends from a cat born in Scotland in 1961 with ears that were folded forward. This cat was a mixed-breed animal of uncertain origin, but the gene causing the fold proved to be dominant, so mating the mutated cat with others resulted in the development of the breed. The Scottish Fold is a medium-sized cat with a broad chest and stocky body. Scottish Folds have a relatively large, round head, with round eyes. A wide variety of coat colors and eye colors are allowed. The ear is rounded at the tip, and ideally is folded directly at the skull (see Figure 16-12).

Siamese (sī-ah-mēz) The Siamese is one of the most ancient breeds of cats, and has been bred for over 200 years. Originally bred in what we now call Thailand, the Siamese was long the preferred cat of the royal family of Siam. Siamese cats have long, lean, muscular, angular bodies (see Figure 16-13). Their face is relatively long, with brilliant blue eyes, and colorpoint markings. The cat is also known for its distinctive vocalization, and its willingness to vocalize to express its opinion.

Sphynx (sfinks) The Sphynx is a rare but distinctive breed of cat. The most distinctive feature is the remarkable lack of hair. The Sphynx is a medium-sized cat with a muscular body, slender legs, and a long thin tail. The Sphynx neck is relatively long, and the head looks large for its body. The ears are large and set wide on the head (see Figure 16-14). This unique breed of cat may have downlike hair on the feet.

Tonkinese (tong-kih-nēz) First established in Canada, the Tonkinese was developed by breeding Siamese and Burmese cats in the 1960s. Tonkinese cats have long legs, a long tail, and a muscular body. They have darker points than the body, but the points and body are less contrasted than in the Siamese. Their hair coat is thick and dense, similar to that of a mink. Tonkinese cats have blue eyes that range from turquoise to aqua in color.

FIGURE 16–14 Sphynx (Courtesy of CFA)

USES

Cats have played many roles since they first began to live in close proximity to humans. Following is a list of current uses of cats in our society:

Companions Cats serve almost exclusively as companions in U.S. society. They are second only to dogs in their popularity, and many people who own cats, own more than one.

Offspring The exhibition of cats is a smaller market than in dogs, but there is some market for the sale of cats as companions or exhibition animals. Because most cats in the United States are of mixed breed, and cats are readily available for adoption, sale of offspring is a very limited market.

Sale of animals As with the sale of offspring, the sale of animals for breeding is very limited in cats when compared to other species.

Work The primary work for cats is hunting for mice and other vermin. Some farms have cats for the purpose of reducing the rodent population. These cats often live out of doors, and are managed as working animals, not as companions.

MANAGEMENT TERMS

***Ad libitum* feeding** (ahd lihb-ih-tuhm) Feed being available all of the time so cats can consume as much as they want. Care must be taken that the cats do not consume too much food and become obese.

Agouti (ah-goo-tē) A hair color that results from alternating stripes of color on the same hair.

Albino (ahl-bī-nho) An animal that completely lacks pigmentation. These animals have white hair, pink skin, and pink or blue eyes. Albino cats are frequently deaf.

Alley cat A homeless cat.

Allogrooming (ăl-lō-grōōm-ing) The practice of grooming other members of the same social group.

Alopecia (ahl-ō-pē-shah) Hair loss that can result from disease or infestation with external parasites.

Altricial (ahl-trish-ahl) Young that are very dependent after birth; kittens are altricial.

Animal shelter A place for animals that do not have a home. Animals may be placed in shelters by the owners, or are found as strays. Most animals in shelters are euthanized due to lack of space.

Awn hairs The secondary hairs below the guard hairs.

Bicolor A hair coat of white and one other color.

Bloom (bloom) The sheen of a coat in top condition.

Brush The fluffy tail of a long-haired cat.

Calico A hair coat of three colors, usually white, black and orange, or variations of those colors. This color pattern occurs almost exclusively in females, and a male calico is extremely rare.

Calling The vocalizations of a female cat in estrus. These vocalizations may occur spontaneously, or in response to stimulation of petting.

Canine teeth (kā-nīn) Long pointed teeth next to the incisors. These are especially prominent in cats that have two on the upper jaw and two on the lower jaw. They are also known as fangs, and are used to hold prey (see Figure 16-15).

Cattery A place where cats are raised or kept.

Cobby Describes a stocky, rounded body style with substantial legs.

Colorpoint A term referring to when the face, paws, and tail are a darker color than the rest of the body (see Figure 16-16). The following are examples of different colorpoints:

Blue point Bluish points.

Chocolate point Brown points.

Lilac point Light grayish points.

Seal point Dark brownish black points.

Dam A female parent.

Declawing The surgical removal of the claws to prevent inappropriate scratching.

Digitigrade (dihg-iht-ih-grād) A cat that walks on its toes.

Digest (dī-jehst) (1) To break down food to a usable form; or (2) A flavor enhancer for cat food that is created by using enzymes to break down animal proteins, which are then sprayed on the food to provide flavor.

Ear mites Mites that live in the ear and consume ear secretions. Ear mites are highly contagious from cat to cat, and are characterized by excessive itching of the ears and dark discharge from the ear. Ear mites can be treated effectively with miticides and cleaning.

Euthanasia (yoo-thah-nā-zha) The humane killing of an animal. Millions of dogs and cats are euthanized in animal shelters annually.

Fault A quality that is counted against a show animal.

Felidae (fē-lih-dā) The order of mammals including cats.

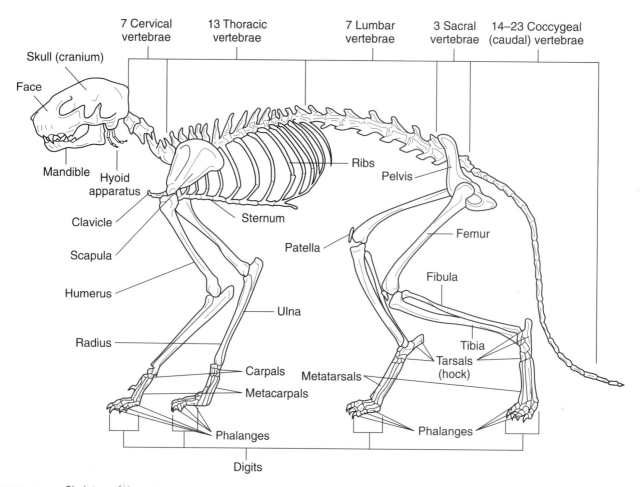

FIGURE 16–15 Skeleton of the cat

FIGURE 16–16 Colorpoint cat

Feral (fair-ahl)   A cat of domestic descent that has returned to a wild state.

Food hopper   A food-storing device that drops food into a dish as it is consumed. This is an especially convenient way to feed cats when people are gone for a few days. Food hoppers work best with dry foods.

Frost   A pinkish gray color.

Guard hairs   Relatively long, coarse hairs on the outside of the coat.

Hairball   A mass of hair that has been ingested and then becomes compacted in the cat's digestive tract. Hairballs may be vomited up.

Household cat   A division at cat shows for nonpedigreed cats.

Judge   A person who determines what cat wins the class at a cat show.

Kitten   A young cat.

Kneading   A repetitive massaging motion of the paws. Kittens often knead the mammary gland while nursing, and some cats knead a location before lying down.

Libido (lih-bē-dō)   The sex drive of an animal.

Limit-feeding   Feeding only a certain amount of food, not necessarily all the animal can consume. Cats are often limit-fed to prevent obesity.

Litter   The kittens born as a result of one pregnancy.

Littermates   The puppies or kittens that are all part of the same litter.

Litter pan (box) A container filled with an absorbent material (**litter**) in which a cat defecates or urinates.

Longy A Manx cat with a tail that approaches full length, but is still disproportionately short for the body. These cats are not eligible to show in the United States.

Lordosis (lōr-dō-sihs) A position in which females in estrus crouch and present their hindquarters to a male.

Mask The coloration around the eyes and face of a cat, as seen in breeds such as the Siamese.

Molt The shedding of a significant portion of hair coat.

Muzzle (muh-zuhl) The lower part of the face including the nose, nostrils, and lower jaw.

Natural breed A breed developed in nature, not by human selection.

Nocturnal (nohck-tər-nahl) Active primarily at night.

Nose pad (nose leather) The skin on the end of the nose around the nostrils.

Obese Excessively fat. Obesity leads to a variety of health problems.

Obligate carnivore An animal that must have a meat-based diet. Cats are obligate carnivores, and must have meat in their diet to meet all of their nutritional needs. Commercial cat foods selected should have meat as the first ingredient.

Odd-eyed A cat with eyes of different colors.

Pedigreed cat A cat that is registered with the CFA, or other breed association.

Phenolic compounds (fē-nohl-ihck) Compounds in cleaning solutions such as Lysol brand that are toxic to cats.

Polydactyl (pohl-ē-dahck-tihl) Having more than five toes on each foot.

Progeny (prah-jehn-ē) The offspring of an animal.

Purring A rumbling sound deep in the chest made by cats.

Queen A female cat.

Queening Delivering of kittens.

Riser A Manx cat with a few tail vertebrae that can be seen or felt (also called a **rumpy-riser**). These cats are eligible to show in the United States.

Rumpy A Manx cat with no tail at all, and with a dimple where the tail would be. These animals are eligible to show in the United States.

Sex-linked trait An inherited trait that is associated with the gender of the cat (for example, a calico coat coloring).

Socialization The exposure of a young animal to a wide variety of stimuli, locations, people, and animals to familiarize them with such exposure.

Spontaneous mutation A genetic change that occurs on its own in one generation.

Stray A cat with no home or owner.

Stubby A Manx cat with a short tail (also called a **stumpy**). These animals are not eligible to show in the United States.

Tabby A hair coat pattern of striped (ticked) hairs and solid hairs.

Ticked The hair pattern in which colors alternate on the hair strand.

Tom A male cat.

Tortoiseshell A hair coat combination of black and orange, or variations of these colors, that are almost always seen in females.

Undercoat The soft, downy hair closest to a cat's skin.

Variety A subcategory within a breed.

Vestibular (vehs-tihb-yu-lahr) **apparatus** The organ in the inner ear of cats that allows them to know where the ground is so they can land on their feet when they fall.

Vet check Prior to a show, the examination of cats to make sure they are healthy and meet the show's health requirements.

Vibrissae (vih-brihs-ah) Cats' whiskers; the long stiff hairs near the muzzle.

Wild cat Any of a variety of cats such as lions, tigers, and bobcats that have not been domesticated. Although some people attempt to keep these animals as "pets," they are very dangerous, and are not suited as pets.

Xenophobia (zehn-ah-fō-bē-ah) The fear of anything novel or new, including strangers. Xenophobia can be a problem for cats that are not socialized.

CHAPTER SUMMARY

Cats have been part of human society for thousands of years. Their primary work-related role has been in the hunting of vermin that are attracted to human food stores. At different times in history, cats have been revered as godlike creatures, and persecuted as harbingers of evil. Cats currently enjoy a place as the second most popular companion animal in the United States. Although a very small percentage of cats in the United States are registered, the Cat Fanciers Association recognizes more than 40 breeds. In the United States, most cats are of mixed-breed descent, and many are feral, and live on the outskirts of human society. Millions of cats are euthanized annually in animal shelters and humane societies across the country. Overpopulation is one of the largest social challenges currently facing the companion animal industry.

STUDY QUESTIONS

Match the cat breed with the characteristic.

1. _____ Abyssinian   a. A cat with a short tail.

2. _____ Egyptian Mau   b. A spotted coat pattern similar to a wild cat.

3. _____ Siamese   c. An ancient breed with short hair and a ticked coat.

4. _____ Burmese   d. A long-haired cat with a round head.

5. _____ Exotic   e. A short-haired cat developed in the eastern United States.

6. _____ Persian   f. A long-haired cat that is large with a straight face.

7. _____ American Shorthair   g. A short-haired cat with a short nose and round head.

8. _____ Ocicat   h. An ancient breed with a spotted coat.

9. _____ Manx   i. A sturdy cat originating in Southeast Asia.

10. _____ Maine Coon   j. A lean-bodied cat with dark points and a distinctive voice.

11. What organization registers and recognizes breeds of cats in the United States?
    a. AFC
    b. CFA
    c. FAC
    d. AKC

12. What breed of cat has no hair?
    a. Siamese
    b. Sphynx
    c. Maine Coon
    d. Exotic

13. What is the most popular registered breed of cat?

14. What two cat colors are sex-linked?

15. What term is used for a female cat?

16. What are the three layers of hair on most cats?

17. On the Internet, find out how many stray cats were euthanized in the United States in the current year. Is that number higher or lower than the previous year?

Chapter 17
Companion Birds

Birds are increasingly popular as pets in the United States. Birds come in a wide variety of colors, sizes, and personalities, and can be excellent companions. Some types of birds sing, others can learn to talk, and some birds can learn more complex tasks. Bird ownership is a large responsibility. Birds can have very specific needs in regard to nutrition, housing, and social interaction, so it is important to thoroughly research bird ownership before acquiring a pet bird. Furthermore, some birds may live up to 100 years, and outlive their owners.

EXTERNAL ANATOMY
OF THE COMPANION BIRD

Although companion birds and poultry have many similarities, they have some important differences. Some of those important differences relate to the external parts of the bird (see Figure 17-1), and the terms that are used to describe those parts. The internal anatomy of companion birds, including the digestive system, skeleton, and reproductive tract, are virtually the same in companion birds and production birds. The following are terms related to the external parts of companion birds:

Beak   The hard mouthpart of a bird. The beak is shaped differently for the various species of birds. Songbirds have pointed beaks whereas parrot-related birds have curved beaks.

Bend of wing   The joint where the humerus meets the ulna and radius.

Breast   The chest of the bird.

Central tail feathers   The primary feathers at the center of the tail.

Cere (sēr)   The thick skin where the beak joins the head of the bird. The cere may be different colors in different sexes of the same species of bird.

Cheek   The area of the face under the eye of a bird.

Coverts (kuhv-erht or kō-vehrt)   The small feathers that cover the bases of large feathers on the body. Covert feathers are named based on their location.

Crown   The top of the head.

Ear coverts   Feathers over the ears.

Forehead   The part of the head in front of the eyes and above the beak.

Foreneck   The area where the wishbone is located, and that is analogous with the cervical vertebrae right above the breastbone.

Lateral tail feathers   The large tail feathers to each side of the central tail feathers.

Mantle   The feathers across the back.

Nape   The back of the neck.

Orbital ring   A ring of unfeathered skin around the eye.

Secondaries   Smaller, lighter feathers than primary feathers.

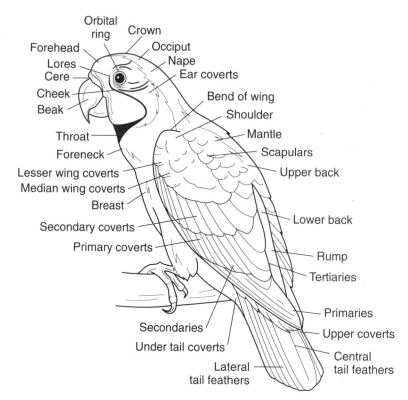

FIGURE 17–1 External anatomy of companion birds

ORDER

Birds are not divided into breeds as are many of the other animals. Two primary orders of birds constitute companion birds. Birds will be identified by the following order and species:

Passerines

Passerine (pahs-er-ēn) is the largest order of birds, with more than 50 percent of bird species being passerines. Most songbirds are passerines. Passerines have three forward-facing toes, and one rear-facing toe. Passerines are also referred to as **Passeriformes** (pahs-ər-ih-fōrmz). The following are birds in the passerine order:

Canary   Canaries originated on the Canary, Madeira, and Azores islands off the coast of Spain, and are members of the finch family. These small birds are most often thought of as yellow, but can be a wide variety of colors. Their lifespan is approximately 15 years.

Finch   A broad category of passerine birds that may be kept as companions. There are many varieties of finches, each with distinctive characteristics. These birds have short, pointed beaks, and may be a variety of colors. Common pet finches include zebra finch, purple finch, goldfinch, and society finch, in addition to the canary that was previously mentioned.

Psittacines

More than 300 species of birds are in the psittacine (siht-ah-sen) order. Many of our common companion birds are in this order. Psittacines have strong, curved beaks, and four toes, with two facing forward, and two facing backward.

African gray parrot   One of the most popular of the parrots, the African gray is primarily grey, with a red tail. The bird is medium-sized (13 inches), and has two common varieties: the Congo gray and the Timneh gray. African grays are very intelligent, and may learn to talk. Some African grays have learned to count and to differentiate colors. They may live more than 50 years.

Amazon parrot   Birds between 10 and 20 inches in length, Amazon parrots have a heavy body and relatively short wings. They originated in South America, and come in a wide variety of brightly colored species.

Budgerigar (parakeet) (buhdg-rē-gahr)   A native of Australia, the "budgie" is a very popular "beginner" bird because of its small size and bright coloring. The birds are about 7 inches in length, and colors include albino, blue, violet, and yellow, as well as the traditional green.

Cockatiel (kohck-ah-tēl)   The cockatiel is native to Australia, and is approximately 12 inches long.

Cockatiels may be gray, white, yellow, or a combination of these colors. Their average lifespan is 25 years.

Cockatoo (kok-ah-tōō) The cockatoo is native to Australia and is primarily white in color. These birds can range between 12 and 25 inches, depending on the variety selected. The most distinctive characteristic is the contrasting-colored crest of feathers on the head of the cockatoo that the bird can raise when excited. Cockatoos can live up to 75 years.

Conures (kahn-yers) Conures originated in Central and South America and can range in size from between 9 and 12 inches. They come in a wide variety of colors.

Eclectus parrots (ehk-lehk-tuhs) Eclectus parrots originated on islands in the South Pacific. They are 14 inches long, and either green with a yellow beak (male), or red with a black beak (female). Eclectus parrots may live up to 30 years.

Lovebird The lovebird is a small species of parrot from Africa. These birds are between 4 and 7 inches in length, and can be a variety of colors and patterns. Lovebirds live from between 5 and 15 years.

Quaker parrot Quaker parrots are also known as **monk parrots,** and originated in South America. They are between 10 and 11 inches long, and are primarily green, although other colors rarely appear. Quaker parrots live between 25 and 30 years.

Parrotlets These tiny parrots are less than 6 inches long, and originated in South America. They are intelligent and can learn tricks. Their expected lifespan is between 20 and 30 years.

USES

Sale of Animals

There is a market for the sale of birds and hatchlings to prospective bird owners, or to retail shops that sell birds.

Work

Historically, some types of birds, such as falcons, were used by the aristocracy for hunting small game; however, birds serve exclusively as companions in modern society.

AVIAN MANAGEMENT TERMS

Aviary (āv-ē-ār-ē) An outdoor enclosure for birds that is large enough for birds to fly freely.

Beak trimming The practice of trimming the tip of the beak to keep the beak properly aligned. The bird may experience difficulty in eating if the beak is not maintained.

Cage A portable enclosure for a bird. A properly sized cage should allow birds to fully extend their wings.

Cuttlebone (kuht-tehl-bōn) The shell of a cuttlefish. The cuttlebone is provided in the cage for the bird to wear its beak down.

Hand-raised A bird that has been raised by humans.

Nail trimming A practice in which the ends of toenails are trimmed. It is important to trim the nails so birds do not scratch themselves, or the owner.

Feather clipping The practice of trimming the primary feathers to limit flight. Feather clipping is an important way of controlling birds.

Feather plucking A behavior in which birds remove their own feathers. Feather plucking is an undesirable behavior, and indicates a stressed bird.

Perch A stick or dowel provided for the bird to sit on in the cage. Some perches have rough surfaces that assist in wearing down toenails.

Self-mutilation Any behavior in which a bird causes injury to itself.

CHAPTER SUMMARY

Birds are the third most popular companion animal in the United States. Companion birds range in lifespan from 5 years to over 50 years, so the commitment to have a companion bird is a major one. Each species has significant differences in nutritional, social, and housing needs, so it is important to research the species before selecting a companion bird.

STUDY QUESTIONS

Match the bird with the characteristic.

1. _____ African gray
2. _____ Canary
3. _____ Cockatoo
4. _____ Monk parrot

a. A passerine.

b. Green South American parrot.

c. A parrot that may learn to talk.

d. An Australian bird with a distinctive crest.

5. What item can be provided in the cage for birds to naturally wear down their beaks?

6. What is the most common order of birds?

7. Name two species of birds that can learn to speak.

8. What is another name for a parakeet?

9. What type of bird was once used for hunting?

10. What management practice limits the ability of the bird to fly?

11. What order of birds has two toes facing forward, and two toes facing backward?

12. List three types of finches that are common pets. Select a species of companion bird, and research the proper diet for that bird.

Chapter 18
Exotic Companion Animals

Chapter Objectives

▶ Understand the challenges of having an exotic companion animal

▶ Learn the variety of exotic companion animals in the United States

People have made pets of a variety of animals over the course of human history. Some of these animals, such as the dog and cat, have adapted so well to domesticity that they have become standards as companion animals. Some people wish for a companion that is unique and out of the ordinary, and turn to exotic animals as companions.

Many challenges are associated with having exotic companion animals, including meeting their social, nutritional, and housing needs. It is important to extensively research the needs of an exotic pet before acquiring one. Very little infrastructure is in place to provide assistance if problems arise with these pets. Finding medical care for exotic pets can also be challenging, so locating a veterinarian that has the experience to treat the animal is important. Many exotic companion animals are tamed wild animals, and are not domesticated. This leads to animals that may have a wide range of behaviors that may not be acceptable in a companion environment.

Choosing to take on the responsibility for any animal should be carefully decided, and this is especially true for exotic companion animals. Something as simple as going out of town for a weekend becomes more complicated with an exotic companion animal because care for the animal will be harder to find than it is for a dog or cat.

VARIETIES OF EXOTIC COMPANION ANIMALS

Most exotic companion animals are identified by species, not breed. Some exotic animals are dangerous, and have been banned by local or state legislation. Before acquiring an exotic animal, ensure that it is legal in the community. Also work with a veterinarian who specializes in exotic pets to ensure that the animal is properly vaccinated. Be aware that even if you vaccinate an exotic animal, it may be treated as an unvaccinated animal if it bites a person, due to the lack of research regarding the effectiveness of vaccines in some exotic animals. This is especially important in regard to rabies, which can be transmitted to humans from any mammal. The following are examples of exotic pets:

Chinchillas (chihn-chihl-ah) A small rodent from South America, the chinchilla was hunted nearly to extinction because of its soft, dense fur. Chinchillas are very active animals with distinctive large ears (see Figure 18-1). They are friendly and rarely bite. They can be any shade of gray, white, or black.

Ferrets (fair-ehts) Ferrets are members of the weasel family, and have been domesticated for hundreds of years. Although now exclusively companion animals, ferrets have been used for hunting throughout history (see Figure 18-2). The ferret has a long, lean body, and is an active and entertaining pet. Ferrets are available in more than 20 color varieties.

FIGURE 18–1 Chinchilla (Courtesy of Joel Sartore/Corbis)

FIGURE 18–3 Gerbil (Courtesy of Ken Usami/Corbis)

FIGURE 18–2 Ferret (Courtesy of GK Hart/Vikki Hart/Corbis)

FIGURE 18–4 Abyssinian cavy (Courtesy of Brittany Bellows)

Frogs Frogs are amphibians and are becoming popular as small pets. The following are popular frog pets:

African clawed frog A small frog that lives exclusively in the water (**aquatic**). The front legs, which have claws, are not strong enough for the frog to be on land.

Bullfrogs A large frog that spends some time in water and some on land. Bullfrogs have loud distinctive voices, and may live over 30 years.

Tree frogs A variety of tree frogs of different colors and patterns are available as pets. Tree frogs are **arboreal** (ar-bōr-ē-əhl), and must have trees and twigs in their environment, as well as moisture.

Gerbils (jər-bəhls) The most common gerbil used as a pet is the Mongolian gerbil, which is from Southeast China and Mongolia. These are small rodents with short hair, long tails, and bright dark eyes (see Figure 18-3). Their most common color is brown, but other color varieties have been developed.

Guinea pigs (gihn-ē) The guinea pig is a short-legged, large-bodied animal with small ears. Guinea pigs originated in South America where they were raised for thousands of years as a meat source. Guinea pigs have short-haired and long-haired varieties, and are bred in a wide variety of colors and patterns. Several breeds of guinea pigs are exhibited.

Guinea pigs are also known as **cavies** (kā-vēz), and are known as such in the exhibition arena. The following are breeds of guinea pigs:

Abyssinian A variety with rosettes in the hair coat (see Figure 18-4).

American The most popular breed of guinea pig, the American has a dense body and a short, straight coat (see Figure 18-5).

Coronet A guinea pig characterized by a long lock of hair on the top of the head (see Figure 18-6).

Peruvian A guinea pig with uniform long hair over the entire body (see Figure 18-7).

Teddy A guinea pig with a short, kinky hair coat (see Figure 18-8).

Hamsters Small rodents that originated in the Middle East, hamsters are popular pets. They can have either long or short hair, and are found in a wide variety of colors. Hamsters are stockier and heavier-bodied than gerbils, with short tails.

Lizards A wide variety of lizards are purchased as companions or pets. Each species has very distinct environmental and nutritional needs. Lizards and other cold-blooded animals need special care in regard to their environment. Because they cannot maintain their own body temperature, they must have cages that can provide supplemental heat. For

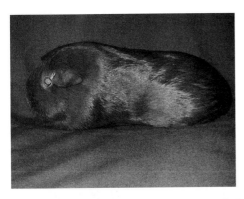

FIGURE 18–5 American cavy (Courtesy of Brittany Bellows)

FIGURE 18–7 Peruvian cavy (Courtesy of Brittany Bellows)

FIGURE 18–6 Coronet cavy (Courtesy of Brittany Bellows)

FIGURE 18–8 Teddy cavy (Courtesy of Brittany Bellows)

some types of lizards, maintaining proper body temperature is important for digestion of their food, and utilization of nutrients. Care must be taken that the environment is not too hot, and that the animal does not become scalded on the heat source. Common species of lizards as pets include the following:

Anole (ah-nōl) A small lizard that is from the same family as the iguana, the anole is between 6 and 18 inches long. Depending on the species, anoles can be green, brown, or gray.

Chameleon (kah-mēl-yhun) Small lizards that have some ability to change color to match their surroundings, chameleons primarily eat insects, which they catch with their sticky tongues.

Dragon A medium-sized lizard with a muscular body, the Australian bearded is a common dragon species and ranges in color from light to dark brown, to bright orange.

Gecko (gehk-ō) Small lizards that are popular as pets, geckos may be a variety of colors and are easily kept in captivity.

Iguana (ih-gwahn-ah) There are more than 600 species of iguana. Green iguanas are the most common companion animals. They grow up to 6.5 feet in length.

Monitor A large lizard that has powerful jaws and claws that can cause significant injury. Monitors are banned in some communities because of the potential for injury. Savannah monitors are the most common pet monitors, and can grow up to 5 feet long, and live up to 15 years.

Mice White mice are most often used as pets. Mice have a short life span of between 2 and 3 years.

Prairie Dogs Prairie dogs are large rodents originally from the American West. Prairie dogs are brown in color, with short dense hair. They are very social animals, and it is recommended that owners buy prairie dogs in pairs.

Rats Varieties of laboratory rats are most often used as pets. These rats can be white or black and white. Rats are intelligent animals, and some people train them to walk on a leash. Rats live between 2 and 3 years, and are prone to cancers, especially later in life.

Snakes A wide variety of snake species are available as companions or pets. Each species has very distinct environmental and nutritional needs. Common species of snakes as pets include the following:

Boas (bō-uh) Snakes generally originating in South or Central America, boas are constrictors, and kill their prey by suffocation. Boas can grow up to 18 feet in length, and are popular as pets because of their wide range of colors and their docile temperaments. Two common pet varieties are the Peruvian red tail and the Mexican red tail boa.

Corn snake Corn snakes are nonvenoumous snakes indigenous to the United States that live from 10 to 15 years. They can be black, or a variety of colors. These are good snakes for the novice owner because of their moderate size (between 3 and 5 feet) and docile temperament.

Pythons (pī-thahn) Pythons are constrictors, and a variety of species are used as pets. The ball python, which grows from 3 to 5 feet in length, is the most popular species as a pet. It is important to research python species before acquiring one, as some species get very large (more than 20 feet long and 200 pounds), and others can have unpredictable temperaments.

Vipers A broad term for a group of venomous snakes including rattlesnakes. Vipers are very dangerous, and are not recommended as pets.

Sugar glider A small marsupial from Australia, the sugar glider is gray or brown with large dark eyes, and dark and white markings on the face. The sugar glider is highly social, and must have either human companionship, or the companionship of other sugar gliders to flourish.

Turtles There are over 200 species of turtles worldwide, and several are kept as pets. Turtles may either be **aquatic,** and spend much of their time in the water, or **terrestrial,** and spend much of their time on land. The following are types of turtles commonly kept as pets:

Box With several varieties, box turtles are indigenous to the United States and have a dark shell that is rounded on top. When frightened, box turtles will pull their legs, tail, and head completely into their shell.

Painted Painted turtles are primarily olive green or black, with red and yellow markings on the ends of their shells. They also have red and yellow stripes on their legs, tail, and head. They are small turtles (between 4 and 9 inches long) and are good pets.

USES

Exotic companion animals are used only as companions.

MANAGEMENT TERMS FOR EXOTIC ANIMALS

Aquatic (ah-kwa-tihck) Living in water.

Agouti (ah-goo-tē) A hair color in hamsters and gerbils that results from bands of color on the hair shaft.

Boar A male guinea pig.

Carapace (kār-ah-pās) The top of a turtle or tortoise shell.

Cavy (kā-vē) A guinea pig.

Constrictor A snake that suffocates its prey by squeezing it.

Hob A male ferret.

Hooded In rats, coloring pattern characterized by color over the head and shoulders that may include a stripe down the back and other color spots on the body.

Hyperestrogenism A serious condition in female ferrets if the estrus period lasts more than one month. Ferrets are induced ovulators, so will stay in estrus until mating occurs, unless hormonal intervention occurs.

Indigenous (ihn-dihj-ah-nahs) Native to a region.

Jill A female ferret.

Kit A baby ferret.

Marsupial (mahr-soop-ē-ahl) A mammal that gives birth to undeveloped young that mature in a pouch on the female.

Sow A female guinea pig.

Terrestrial (teh-rehs-trē-ehl) Living on land.

Tortoise A type of turtle that lives on land.

Turtle Any of a variety of reptiles with hard shells and the ability to withdraw their head, neck, and tail into their shells.

Venomous An animal that uses venom, or poison, to kill prey.

Viper A type of venomous snake.

CHAPTER SUMMARY

The decision to have an exotic companion animal requires much thought and consideration. A wide range of animals have been selected over the years to serve a role as companions. A prospective exotic pet should be thoroughly researched to ensure you know its nutritional, behavioral, and housing needs. For many species, these needs are not well researched, and commercially available products are not readily available. Veterinary care may also be difficult to find, and research should be conducted to determine the zoonotic diseases that different exotic animals may carry. Finally, it is important to know the zoning laws in a particular area, as some exotic pets are not allowed in certain areas.

STUDY QUESTIONS

Match the animal in the left column with the term in the right column.

1. _____ Boa
2. _____ Guinea pig
3. _____ Sugar glider
4. _____ Anole
5. _____ Viper
6. _____ Zoonotic disease
7. _____ Monitor
8. _____ Jill
9. _____ Gerbil
10. _____ Rat

a. Marsupial
b. Poisonous snake
c. Transferred from animal to human
d. Constrictor snake
e. Large lizard
f. Hooded
g. Mongolian
h. Cavy
i. Female ferret
j. Small lizard

11. What disease can female ferrets get if they do not ovulate?
    a. Pyometra
    b. Hyperestrogenism
    c. Salmonella
    d. Cushing's disease

12. What term describes a species that is native to an area?
    a. Exogenous
    b. Endogenous
    c. Primogenous
    d. Indigenous

13. How would you tell a gerbil from a hamster?

14. What is an aquatic turtle?

15. What is an agouti color?

16. Select an exotic species you would consider as a pet. Research its nutritional and medical needs, and determine if they may have any zoonotic diseases that may be a threat.

Answer Key

CHAPTER 1

1. Rank the species below in the order in which they were domesticated (1 is longest ago).

   __2__ Ovine   __3__ Caprine   __1__ Canine   __4__ Feline

   __3__ Bovine   __5__ Equine   __3__ Porcine

2. List one aspect of animal science that has affected your life today.

   Variable answers. Any meal that contained meat or dairy products. Any interactions with an animal at home.

3. Calculate the number of dogs and cats owned in your community.

   a) Calculate the number of households in your community by dividing the population of your community by the average number of people in a household (2.67 according to the 2000 U.S. Census).

   b) Multiply the estimation formula for the number of dogs (.58) by the total number of households.

   c) Multiply the estimation formula for the number of cats (.66) by the total number of households.

   Example: community of 10,000 people.

   10,000/2.67 = 3,745.32 households
   3,745.32 × .58 = 2,172 dogs
   3,745.32 × .66 = 2,472 cats

   Ensure that students complete the first calculation from population to number of households.

   Please note, these formulas are based on the national averages regarding number of households with pets, and are not designed to reflect the exact number of animals in your community. However, the calculation will provide a reasonable estimate of the number of dogs and cats in your community.

Match the animal science specialization with its description.

5. __e__ Physiology

6. __c__ Nutrition

7. __a__ Ethology

8. __b__ Biotechnology

9. __f__ Reproduction

10. __d__ Management

a. The scientific study of animal behavior as it relates to its environment.

b. Any technique that uses living organisms to make or modify products, to improve plants or animals, or to develop microorganisms for specific purposes.

c. The science of how food is used by the body.

d. The direction or supervision of an animal science entity, which could be a business or the direction of the care of animals.

e. The science of the vital physical functions of living things.

f. The science of how animals produce offspring.

Match the species name with the common name of these animals.

11. __c__ Porcine   a. Goat

12. __f__ Canine   b. Horse

13. __g__ Feline   c. Pig

14. __h__ Bovine   d. Sheep

15. __b__ Equine   e. Bird

16. __d__ Ovine   f. Dog

17. __a__ Caprine   g. Cat

18. __e__ Avian   h. Cow

CHAPTER 2

Match the aspect of the animal science industry with its definition.

1. __c__ Aquaculture   a. A place where beef cattle are fed to slaughter weight.

2. __d__ Boarding stable   b. A place where chicks can be purchased.

3. __j__ Fingerling   c. Production of fish for human consumption.

4. __f__ Stocker   d. A place where an owner can pay a fee to have a horse cared for.

5. __b__ Hatchery   e. An operation that feed pigs until they are ready for processing.

6. __i__ Vertical integration   f. Calves that are postweaning, but too small for the feedlot.

7. __a__ Feedlot   g. A facility that houses dogs.

8. __k__ Seedstock   h. A swine facility that raises pigs from birth to market.

9. __l__ Puppy mill   i. The ownership by one entity of all aspects of production.

10. __g__ Kennel   j. A small fish.

11. __h__ Farrow-to-finish   k. Animals that are used for breeding.

12. __e__ Finishing operation   l. An operation that raises large numbers of dogs in unacceptable conditions.

13. List the operations involved in beef cattle production beginning with the birth of a calf and ending with the consumer purchase of a product.

Seedstock > cow-calf > feedlot > slaughter

              stocker

14. What are the primary products of animal agriculture?

Milk, meat, fiber, offspring

15. List 10 careers in animal agriculture that are not listed in this chapter. Include the degree of education required or recommended for each position.

Variable answers. Objective is for students to understand what a wide range of career opportunities are available for those interested in animal agriculture.

CHAPTER 3

1. Complete the following table with the correct information:

Species	Male Castrate	Male Intact
Bovine	Steer	Bull
Canine	Dog	Dog
Equine	Gelding	Stallion
Chicken	Capon or caponette, depending on method	Rooster

Match the stage of the estrous cycle with its description.

2. __d__ Estrus

3. __c__ Proestrus

4. __e__ Metestrus

5. __b__ Diestrus

6. __a__ Anestrus

a. A period when no activity is occurring.

b. The period between estrus periods.

c. The time when FSH is stimulating follicular growth.

d. The period when the female is receptive to the male.

e. The time immediately after the cessation of receptive behavior.

7. Which of the following refers to the change in genes due to natural selection?

a. Population genetics

b. Genetic drift

c. Genetic waves

d. Gene frequency

8. Which describes the visual appearance of the animal?

a. Phenotype

b. Genotype

c. Heterozygous

d. Homozygous

9. Which breeding strategy results in animals that are the most similar genetically?

a. Crossbreeding

b. Natural selection

c. Inbreeding

d. Outcrossing

10. A *dilution* gene affects the expression of a color. This gene changes a *chestnut* horse to a palomino.

11. An allele that is present but not expressed is *recessive*.

12. A male animal with one testicle that does not descend into the scrotum is a/an *cryptorchid* .

13. The surgical procedure to remove all reproductive organs of a female is a/an *ovariohysterectomy*.

14. A horse breeder bred a bay stallion and a bay mare, and they had a chestnut offspring. In horses, the chestnut color is recessive. If bay is represented by B, and chestnut is represented by b, what is the genotype of the foal? What is the genotype of each parent? What percentage of the offspring of this mating would be bay? What percentage would be chestnut?

Foal: bb
Stallion: Bb   Mare: Bb
75% bay offspring; 25% chestnut offspring

	B	b
B	BB	Bb
b	bB	bb

15. Animal species have a wide range of heritable diseases. Select the species of your choice, and conduct an Internet search to determine what heritable diseases are of concern in that species. Select one disease, research how it is inherited, and make a recommendation on how to control the disease in the population.

    Highly variable. Would suggest having students present their findings to class to increase everyone's understanding of genetic diseases.

CHAPTER 4

Match the correct skeletal part with the correct external part.

	Skeletal part		External part
1.	d	Humerus	a. Stifle
2.	f	Cranium	b. Coffin bone
3.	b	P3	c. Croup
4.	g	Third metacarpal	d. Arm
5.	h	Thoracic vertebra	e. Hock
6.	i	Carpal joint	f. Skull
7.	e	Tarsal joint	g. Cannon bone
8.	a	Patellar joint	h. Back
9.	c	Sacral vertebrae	i. Knee

10. Where in the digestive tract does the most absorption of nutrients take place?

    a. Stomach

    b. Small intestine

    c. Large intestine

    d. Cecum

11. Describe the structural differences between a vein and an artery.

    Arteries have three layers of tissue: connective tissue, muscular tissue, and endothelial tissue. Arteries have no valves.

    Veins: less muscular, have valves

12. What four parts make up the ruminant stomach, and what is the role of each part?

    Abomasum: True stomach that digests food and adds gastric juices

    Omasum: Muscular portion that mechanically digests and removes excess water

    Rumen: Largest portion of the ruminant stomach; the site of fermentation

    Reticulum: Honeycomb-like portion of the stomach that contains liquid portions of food. Debris in food often ends up trapped in the reticulum (hardware disease).

13. List the three muscle types, and give an example of where in the body each muscle type is found.

    Striated involuntary muscle: The cardiac muscle found in the heart.

    Striated voluntary muscle: The skeletal muscle that is connected to the skeleton and moves bones.

    Unstriated involuntary muscle: These surround internal organs and work independent of conscious thought.

14. List the types of connective tissue, and give an example of where each type of connective tissue would be found.

   Bone: The skeleton

   Cartilage: In joints between bones, or forms parts independent of bone (for example, ear, nose)

   Ligament: Connects bone to bone, found in any joint

   Tendon: Connects muscle to bone; found wherever skeletal muscle exists

15. What are the parts of the pulmonary circulatory system?

   Lungs, heart, pulmonary artery, pulmonary vein

16. What types of feathers are found on the wings of birds?

   Coverts

17. Properly label the parts of the heart indicated in the following diagram:

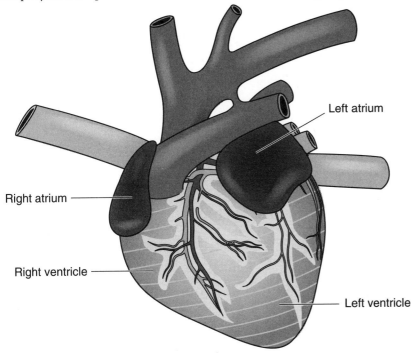

18. Label the parts of the sperm cell, and identify the primary role of each part.

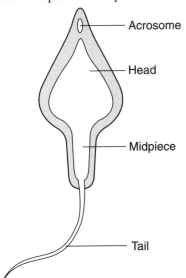

19. List the parts of the female mammalian reproductive tract, in order, from the ovary to the vulva.

    Ovary > infundibulum > oviduct > uterine horn > uterine body > cervix > vagina > vulva

20. List the bones in the front leg of the cow in order from the ground to the top.

    Phalanges (P3, P2, P1) > metacarpus > carpus > ulna and radius > humerus > scapula

21. Name two parts of the avian female reproductive tract that are not found in the mammalian reproductive tract.

    Cloaca, vent, sperm nests

22. What is the purpose of "sperm nests" and in what species do they exist?

    In chickens, they store sperm in the reproduction tract until ovulation occurs, and the ova can be fertilized.

CHAPTER 5

1. What are the five classes of nutrients?

   Carbohydrates, fats, proteins, vitamins, minerals

2. In what part of the digestive tract is pepsin added to the digesta?

   Stomach

3. In what part of the digestive tract do horses ferment roughages?

   Cecum

4. *Legumes* are the forage source with the highest protein.

5. *Maintenance* is the classification for energy need on which all other classifications are based.

6. List three feedstuffs that qualify as concentrates.

   Any feed product that is high in energy and low in fiber: barley, oats, corn, wheat, millet, etc.

7. What nutrient is the most energy dense?

   Fat

8. What part of the ruminant stomach is the site of bacterial fermentation?

   a. Rumen

   b. Reticulum

   c. Omasum

   d. Abomasum

9. What is a balanced ration?

   A ration that meets all of the nutritional needs of the animal, without feeding excess of some nutrients.

10. Call a local feed provider and learn the cost of some common feedstuffs. What class of feeds is most expensive per pound, forages or concentrates?

    Generally concentrates, but can vary if the supply of forages is low

11. Select a species of interest and research what amino acids are essential, and nonessential for that species.

    Variable answer depending on species selected

CHAPTER 6

1. What are the major approaches to the study of animal behavior?

   Behavioral ecology, comparative psychology, ethology, sociobiology

2. What is a stereotypy? Give an example.

   A repetitive behavior with no apparent purpose. Examples include weaving, cribbing in horses, tongue-rolling in hogs, excessive licking in dogs and cats, or pacing in any species.

3. How does successive approximation modify behavior?

   Rewarding increments of the behavior as it approaches the desired behavior.

4. Why is it difficult to modify behavior successfully using punishment?

   - *Timing must be precise for an animal to associate punishment with the behavior.*
   - *The punishment must happen every time the behavior occurs.*
   - *Punishment needs to be associated with the behavior, not the handler.*

5. What can you learn from an animal's body language?

   The primary method of communication from animals is through body language. You can learn how animals respond to their environment at a given time, anticipate potential reactions to a situation, and gain information on the overall well-being of the animals.

6. What message is a cat sending when its ears are flattened to its head?

   It is angry/upset.

7. What is anorexia?

   Failure to eat

8. On the Internet, find pictures of a species of your choice exhibiting different body language. Print the pictures and describe the message the animal is sending.

9. Make an ethogram of your day.

   Example:
   7:00 a.m. Woke up and showered
   7:30 a.m. Ate breakfast
   7:45 a.m. Drove to work
   8:00 a.m.–10:30 a.m. Sat at desk working on X
   10:30 a.m. Walked down the hall to talk to Sam
   Etc.

CHAPTER 7

1. What disease affects the central nervous system and is a zoonotic disease that affects all mammals?

   Rabies

2. Bacteria and viruses are examples of what type of disease-causing organisms?

   Pathogens

3. What is the process of introducing an antigen to the body so that the immune system will respond with antibodies?

   Vaccination

4. What is antibiotic resistance, and why is it a concern in treatment of animal disease?

   Pathogens mutate and develop a resistance to antibiotics, so the antibiotics are no longer of use, or are of limited use, in treating disease. It becomes a concern because animals, and also humans, with disease are more difficult to treat when pathogens develop antibiotic resistance.

5. What disease is the degeneration of a bone in the foot of the horse?

   Navicular disease

6. What ectoparasite causes mange?

   Mites: demodectic and sarcoptic

7. What endoparasite is carried by mosquitoes and affects both dogs and cats?

   Heartworm

8. What management steps can be taken to decrease the incidence of disease?

   Sanitation, proper vaccination, good nutrition

9. What is the difference between a modified-live and a killed vaccine?

   A modified-live vaccine contains a pathogen that has been changed in such a way that it cannot multiply in the animal to reach an infective level, but can still elicit a response from the immune system. A killed vaccine involves a pathogen that is the disease-causing form, but it is killed. The recommendation of which form of vaccine to use varies with the disease and the animal. For example, killed vaccines are often preferred for pregnant females.

10. What internal parasite has the largest impact on the swine industry?

    Ascarid

11. What is the name of the group of chemicals that are used to treat internal parasites in animals?

    Anthelminitics

12. List three diseases that affect the central nervous system, and what species in which they are found.

    Bovine spongiform encephalitis: cattle
    Canine distemper: dogs
    Equine protozoal myelitis: horses
    Scrapie: sheep and goats
    Rabies: all mammals
    West Nile encephalitis: humans, horses, dogs

13. What is a vector?

    A living thing that carries disease from one animal to another. Insects are common vectors.

14. What disease, and in what species, do many states have a law requiring vaccination.

    Rabies in dogs

15. Select one of the diseases listed in this chapter. Research and write a paper on the disease, including species affected, causes, symptoms, and treatment. What management practices can be used to prevent the disease?

CHAPTER 8

Match the breed with the place of origin. Place of origin may be used more than once.

1. __a__ Salers   a. France

2. __b__ Hereford   b. England

3. __c__ Angus   c. Scotland

4. __j__ Santa Gertrudis   d. Germany

5. __a__ Limousin   e. India

6. __i__ Chianina   f. United States

7. __e__ Zebu   g. Australia

8. __f__ Beefmaster   h. Switzerland

9. __h__ Simmental   i. Italy

10. __g__ Murray Grey   j. Texas

11. Which of the following breeds is used extensively in crossbreeding to increase heat and disease tolerance?

    a. Hereford

    b. Brangus

    c. Brahman

    d. Simmental

12. What beef breed also has a dairy breed associated with it?

    a. Hereford

    b. Shorthorn

    c. Murray Grey

    d. Angus

13. The following crossbred is a result of breeding Angus and Herefords.

    a. Beefmaster

    b. Brangus

    c. Simmental

    d. Black baldie

14. What term describes an animal's skeletal size?

    a. Frame

    b. Stature

    c. Mass

    d. Conformation

15. *Creep feeding* is the method of feeding young calves that prevents older animals from accessing the feed.

16. List three beef breeds that are composite breeds, and the breeds that were combined to develop them.

Beefmaster: Hereford/Shorthorn/Brahman
Braford: Brahman/Hereford
Brangus: Brahman/Angus
Santa Gertrudis: Brahman/Shorthorn

17. List the primal, or wholesale, cuts of beef.

Brisket, chuck, flank, loin, plate, rib, round

18. From what primal cut do we get T-bone and porterhouse steaks?

Loin

19. What is the current price per pound that producers are receiving for beef? Compare to the current cost per pound for beef in the grocery store

    Go to http://beef.unl.edu/agReportsMenu.shtml. Select current Ag Prices, then select a sales location in your region.

CHAPTER 9

Match the dairy breed with its characteristic

1. __d__ Jersey
2. __e__ Ayrshire
3. __f__ Brown Swiss
4. __a__ Holstein
5. __c__ Milking Shorthorn
6. __b__ Guernsey

a. Produces the largest volume of milk.
b. Golden brown and white in color.
c. Red and white with roaning acceptable.
d. Produces the highest percentage of butterfat.
e. Originally from Scotland.
f. Bred as a dual-purpose milking and draft animal.

7. What state has the most dairy cattle?
   a. Wisconsin
   b. Texas
   c. California
   d. New York

8. What state produces the most cheese and butter?
   a. Wisconsin
   b. Texas
   c. California
   d. New York

9. Which is used as a house for young calves?
   a. Parlor
   b. Tie stall
   c. Free stall
   d. Hutch

10. What is the DHIA, and what does it do?

    The Dairy Herd Improvement Association collects and maintains production records, and makes them available to member producers to assist in optimal management.

11. What is the purpose of teat dip?

    To disinfect the teat following milking

12. List three dairy products other than fluid milk.

    Butter, cheese, sour cream, half-and-half, evaporated milk, powdered milk, ice cream, cream

13. What is the current price that producers receive for fluid milk in your state?

    http://www.ams.usda.gov/dairy/mncs/index.htm

14. What is the difference between ice milk and ice cream?

    The amount of milk fat present in the product

15. Obtain a container of cream from the grocery store. Pour the cream into a jar with a tight lid. Make sure the jar is large enough that the cream fills it between half and three-quarters full. Shake the cream until it divides into a solid and liquid portion. What is each portion? *Butter and buttermilk*

   If one person uses heavy cream and another uses light cream, which will produce more of the solid product? Why? Heavy cream would make more butter because it has more butterfat present.

   This can also be done with a blender or mixer. Just blend or mix through whipped cream stage until separation occurs.

CHAPTER 10

Match the breed on the left with its characteristic on the right.

1. __i__ Tamworth   a. Black and white with a characteristic belt.
2. __h__ Berkshire   b. Developed in Ohio and involved in the development of several other breeds.
3. __j__ Duroc   c. Large white hogs known for their mothering ability and floppy ears.
4. __a__ Hampshire   d. A Belgian breed with extremely lean carcasses.
5. __f__ Landrace   e. Developed in Indiana with contributions from a breed from Gloucester.
6. __b__ Poland China   f. The premier bacon breed of Denmark.
7. __c__ Yorkshire   g. A red pig with white face and legs.
8. __e__ Spotted Swine   h. The pig that had the first organized registry.
9. __g__ Hereford   i. The "bacon pig" of Ireland and England.
10. __d__ Pietrain   j. A large red pig developed in the eastern United States.

11. A mating that results in a pig that is not going to be used for breeding is called:

    a. Mismating

    b. Market mating

    c. Terminal cross

    d. Dead-end mating

12. What continent is known for breeds of hogs that produce extremely large litters, and reach puberty at an early age?

    a. Asia

    b. Africa

    c. Australia

    d. Europe

13. What small sharp teeth are clipped on baby pigs?

    a. Incisors

    b. Fangs

    c. Needle teeth

    d. Molars

14. What is farrowing?

    a. The mating of two pigs

    b. The process of removing part of the tail to prevent chewing

    c. Feeding pigs in a natural setting

    d. The birth/delivery of baby pigs

15. What is the national organization that promotes the swine industry?

    National Pork Board

16. List three management phases for a hog being prepared for market.

    Growing phase, finishing phase, nursery phase

17. What is vertical integration?

    Ownership by a single entity (individual or company) of animal production, processing, and distribution aspects. Integration may include two or three aspects, or all aspects of production, processing, and distribution.

CHAPTER 11

Match the breed of bird with the class to which it belongs. One of the listed classes is used twice.

1. __g__ Pekin   a. Mediterranean
2. __h__ Khaki Campbell   b. Bantam duck
3. __b__ Call   c. Heavyweight goose
4. __a__ Leghorn   d. Continental
5. __i__ Orpington   e. All other standard
6. __c__ Embden   f. American
7. __f__ Rhode Island Red   g. Heavyweight duck
8. __i__ Old English Game   h. Lightweight duck
9. __d__ Lakenvelder   i. English
10. __e__ Frizzle

11. Which of the following is a variety?

    a. Embden

    b. Bantam

    c. Buff

    d. Orpington

12. Which practice is used to reduce damage done through cannibalization?

    a. Candling

    b. Debeaking

    c. Plucking

    d. Molting

13. Which breed constitutes virtually all commercial laying hens?

    a. Rhode Island White

    b. Wyandotte

    c. Ameracauna

    d. Leghorn

14. What organization publishes the *Poultry Standard of Perfection*?

    American Poultry Association

15. How many classes of geese are recognized in the *Standard of Perfection*, and what are they?

    3: heavyweight, medium weight, lightweight

16. What English class breed is used extensively in the production of commercial broilers?

    Cornish

17. What is a bantam?

    A very small bird

18. What breed of duck is popular for meat production in England?

    Aylesbury

19. Draw and label the primary parts of the egg.

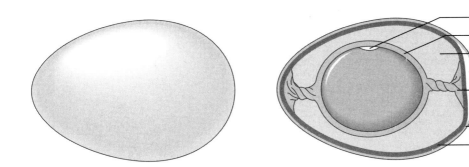

Germinal disc
Vitelline membrane
Albumen or White
Air cell
Chalaza
Shell
Outer membrane
Inner membrane

20. The poultry industry is a common target for people opposed to commercial farming. Identify a controversial topic in the poultry industry, research both sides of the controversy, and prepare a short report for your classmates.

CHAPTER 12

Match the breed of sheep or goat with the description.

1. __h__ Lincoln
2. __i__ Angora
3. __f__ Nubian
4. __b__ Suffolk
5. __c__ Katahdin
6. __d__ Merino
7. __j__ Saanen
8. __e__ LaMancha
9. __g__ Hampshire
10. __a__ Pygmy goat

a. Kept primarily as a pet.

b. The most popular sheep breed in the United States.

c. A hair sheep.

d. An old breed of sheep that was integral in the development of many other breeds.

e. A goat with very small ears.

f. A goat with large, floppy ears.

g. A large sheep with some wool on the face and legs.

h. The largest breed of sheep.

i. A goat raised for its hair.

j. A white or cream-colored dairy goat.

11. Which breed of goat is raised primarily for meat production?

    a. French Alpine

    b. Boer

    c. Toggenburg

    d. Angora

12. What device can be used to castrate lambs?

    a. Elastrator

    b. Browser

    c. Spinner

    d. Tilt table

13. What is the term for a baby goat?

    a. Lamb

    b. Calf

    c. Kid

    d. Wether

14. What are the wool types by which sheep breeds are categorized?

    Carpet wool, fur sheep, hair sheep, long wool, medium wool, fine wool

15. What practice involves feeding females more to increase reproductive efficiency?

    Flushing

16. What are the market classifications of lambs?

    Fed lambs, hothouse lambs, spring lambs, lambs

17. What is a wether?

    Castrated male sheep

18. What breed of sheep sets the standard for wool quality?

    Merino

19. What is culling?

    Permanent removal of an animal from the herd

CHAPTER 13

Match the product with the species that produces it. Although more than one species may produce the product, use each product only once.

1. __b__ Velvet   a. Deer

2. __a__ Venison   b. Elk

3. __e__ Feathers   c. Bison

4. __c__ Mounted heads with humps   d. Rabbit

5. __d__ White meat   e. Ostriches

6. Which is the result of mating a bison with a beef cow?

    a. Cuffalo

    b. Beefalo

    c. Biscow

    d. Cowson

7. Which ratite is not used for production in the United States?

   a. Ostrich

   b. Rhea

   c. Kiwi

   d. Emu

8. What is the form of aquaculture where fish are released into the wild to be recaught?

   a. Food fishing

   b. Mussel fishing

   c. Release fishing

   d. Sport fishing

9. What is the difference between a four-class and six-class breed of rabbit?

   The number of classifications for them at a rabbit show

10. What products can be sold from ostriches?

    Meat, eggs, skin, feathers

11. What is aquaculture?

    The production of fish and fish products for human consumption

12. List three species of fish raised for human consumption.

    Catfish, salmon, trout, tilapia, perch

13. Name one market, other than human consumption, for fish raised in aquaculture production.

    Sport fishing, baitfish

14. What is the largest deer raised?

    Elk

15. What is the difference between a llama and an alpaca?

    Llamas are larger, and have ears that are more banana-shaped, whereas alpacas have smaller, sharper ears and muzzles. Alpacas also have a finer fiber of higher quality.

CHAPTER 14

Match the breed with the fact provided.

1. __e__ Arabian
   a. Best known for racing on the flat.

2. __h__ Percheron
   b. Results when a horse and donkey are bred.

3. __j__ Quarter horse
   c. The smallest of the pony breeds.

4. __g__ Appaloosa
   d. Naturally performs the running walk.

5. __b__ Mule
   e. The oldest and purest breed.

6. __c__ Shetland
   f. A golden draft horse.

7. __i__ Saddlebred
   g. A stock-type horses with a distinctive spotted coat.

8. __d__ Tennessee Walking Horse
   h. A gray or black draft horse.

9. __f__ Belgian
   i. The "peacock of the show ring."

10. __a__ Thoroughbred
    j. The most popular breed in the United States.

11. What rodeo event involves jumping off the horse to tackle a steer?

    a. Calf roping

    b. Steer wrestling

    c. Cow tipping

    d. Steerdogging

12. How many barrels are in a barrel racing pattern?

    a. 4

    b. 3

    c. 2

    d. 1

13. What three horses were the foundation sires for the Thoroughbred?

    Darley Arabian, Byerly Turk, Godolphin Arabian

14. What organization is the legislative arm of the horse industry?

    American Horse Council

15. What color is a palomino?

    Golden yellow with a white mane and tail

16. What Indian tribe created the Appaloosa horse?

    Nez Perce

17. What breed races with a sulky?

    Standardbred

18. What three races are in the Thoroughbred Triple Crown?

    Kentucky Derby, Preakness Stakes, Belmont Stakes

19. Draw and label three leg markings.

    Refer to Figure 14-7

CHAPTER 15

Match the breed of dog with the group it belongs in.

1. __c__ Ibizan   a. Terrier

2. __d__ Great Pyrenees   b. Herding

3. __e__ Shih tzu   c. Hound

4. __g__ Dalmatian   d. Working

5. __f__ English Setter   e. Toy

6. __a__ Airedale terrier   f. Sporting

7. __b__ Australian shepherd   g. Nonsporting

8. What herding breed has long curly hair, similar to dreadlocks?

   a. Puli

   b. Vizsla

   c. German shepherd

   d. Greyhound

9. What sport involves dogs running over a course of jumps and obstacles?

   a. Flyball

   b. Lure coursing

   c. Bench trial

   d. Agility

10. What are the two breeds of corgis?

    Pembroke and Cardigan

11. What is the difference between a sight hound and a scent hound?

    Sight hounds are lighter-framed dogs that use vision and speed to track and chase prey. Scent hounds are heavier-bodied dogs, typically with long, drooping ears, that use scent to track and locate prey.

12. List three organizations that train and place service dogs.

    Canine Companions for Independence, the Delta Society, Leader Dogs for the Blind

13. What organization certifies boarding kennels?

    American Boarding Kennel Association

14. What is an undershot jaw?

    The lower jaw projects out farther than the upper jaw. This is a characteristic in some breeds, and a major fault in others.

15. What is the miscellaneous group? Go to the Internet and find what breeds are currently in the miscellaneous group.

    The miscellaneous group is comprised of dogs that are on the waiting list for full admission into the American Kennel Club.

CHAPTER 16

Match the cat breed with the characteristic.

1.	c	Abyssinian	a. A cat with a short tail.
2.	h	Egyptian Mau	b. A spotted coat pattern similar to a wild cat.
3.	j	Siamese	c. An ancient breed with short hair and a ticked coat.
4.	i	Burmese	d. A long-haired cat with a round head.
5.	g	Exotic	e. A short-haired cat developed in the eastern United States.
6.	d	Persian	f. A long-haired cat that is large with a straight face.
7.	e	American Shorthair	g. A short-haired cat with a short nose and round head.
8.	b	Ocicat	h. An ancient breed with spotted coat.
9.	a	Manx	i. A sturdy cat originating in Southeast Asia.
10.	f	Maine Coon	j. A lean-bodied cat with dark points and distinctive voice.

11. What organization registers and recognizes breeds of cats in the United States?

    a. AFC

    b. CFA

    c. FAC

    d. AKC

12. What breed of cat has no hair?

    a. Siamese

    b. Sphynx

    c. Maine coon

    d. Exotic

13. What is the most popular registered breed of cat?

    Persian

14. What two cat colors are sex-linked?

    Calico and tortoiseshell

15. What term is used for a female cat?

    Queen

16. What are the three layers of hair on most cats?

    Undercoat, awn hairs, guard hairs

17. On the Internet, find out how many stray cats were euthanized in the United States in the current year. Is that number higher or lower than the previous year?

CHAPTER 17

Match the bird with the characteristic.

1. __c__ African gray   a. A passerine.

2. __a__ Canary   b. Green South American parrot.

3. __d__ Cockatoo   c. A parrot that may learn to talk.

4. __b__ Monk parakeet   d. An Australian bird with a distinctive crest.

5. What item can be provided in the cage for birds to naturally wear down their beaks?

    Cuttlebone

6. What is the most common order of birds?

    Passerine (Passeriformes)

7. Name two species of birds that can learn to speak.

    African gray, budgerigar

8. What is another name for a parakeet?

    Budgerigar

9. What type of bird was once used for hunting?

    Falcon

10. What management practice limits the ability of the bird to fly?

    Wing clipping

11. What order of birds has two toes facing forward, and two toes facing backward?

    Psittacine

12. List three types of finches that are common pets. Select a species of companion bird, and research the proper diet for that bird.

    Canary, zebra finch, purple finch, goldfinch, society finch

CHAPTER 18

Match the animal in the left column with the term in the right column.

1. __d__ Boa
2. __h__ Guinea pig
3. __a__ Sugar glider
4. __j__ Anole
5. __b__ Viper
6. __c__ Zoonotic disease
7. __e__ Monitor
8. __i__ Jill
9. __g__ Gerbil
10. __f__ Rat

a. Marsupial
b. Poisonous snake
c. Transferred from animal to human
d. Constrictor snake
e. Large lizard
f. Hooded
g. Mongolian
h. Cavy
i. Female ferret
j. Small lizard

11. What disease can female ferrets get if they do not ovulate?

    a. Pyometra

    b. Hyperestrogenism

    c. Salmonella

    d. Cushing's disease

12. What term describes a species that is native to an area?

    a. Exogenous

    b. Endogenous

    c. Primogenous

    d. Indigenous

13. How would you tell a gerbil from a hamster?

    Gerbils have longer tails, smaller frames, and usually shorter hair.

14. What is an aquatic turtle?

    A turtle that lives in water

15. What is an agouti color?

    Color resulting from banding on the hair shaft

16. Select an exotic species you would consider as a pet. Research its nutritional and medical needs, and determine if they may have any zoonotic diseases that may be a threat.

REFERENCES

American Standard of Perfection 1998. Published by the American Poultry Association, Inc. Mendon, Mass.

Campbell, K. L., Corbin, J. E., and J. R. Campbell. *Companion Animals: Their Biology, Care, Health and Management.* Pearson Prentice Hall. Upper Saddle River, NJ.: 2005.

Case, L. P. *The Cat: Its Behavior, Nutrition, and Health.* Iowa State Press: 2003.

Case, L. P. *The Dog: Its Behavior, Nutrition, and Health.* Iowa State Press: 1999.

Damron, W. S. *Introduction to Animal Science: Global, Biological, Social and Industry Perspectives.* 3rd ed. Pearson Prentice Hall. Upper Saddle River, NJ: 2006.

Gillespie, J. R. *Modern Livestock and Poultry Production.* 7th ed. Delmar Learning. Clifton Park, NY: 2004.

Herron, R., and J. A. Romich. *Delmar's Veterinary Technician Dictionary.* Delmar Learning. Clifton Park, NY: 2000.

Parker, R. O. *Equine Science.* 2nd ed. Delmar Learning. Clifton Park, NY: 2003.

Pollard, M. *The Encyclopedia of the Cat.* Parragon Publishing. Bath, UK: 2003.

Romich, J. A. An *Illustrated Guide to Veterinary Medical Terminology.* 2nd ed. Delmar Learning. Clifton Park, NY: 2006.

Romich, J. A. An *Illustrated Guide to Veterinary Medical Terminology.* 1st ed. Delmar Learning. Clifton Park, NY: 2000.

Taggart, C. (ed). *Encyclopedia of the Dog.* Octopus Publishing Group, Ltd. London, UK: 2000.

Webster Dictionary and Thesaurus. Deluxe Edition. Nichols Publishing Group. 2004.

Index